Columbia University

Contributions to Education

Teachers College Series

No. 58

AMS PRESS
NEW YORK

PRACTICE IN THE CASE OF SCHOOL CHILDREN

BY

THOMAS JOSEPH KIRBY, Ph.D.

PROFESSOR OF ELEMENTARY EDUCATION AT THE UNIVERSITY OF PITTSBURGH

TEACHERS COLLEGE, COLUMBIA UNIVERSITY
CONTRIBUTIONS TO EDUCATION, NO. 58

PUBLISHED BY
Teachers College, Columbia University
NEW YORK CITY
1913

Library of Congress Cataloging in Publication Data

Kirby, Thomas Joseph, 1877-
Practice in the case of school children.

Reprint of the 1913 ed., issued in series: Teachers College, Columbia University. Contributions to education, no. 58.
Originally presented as the author's thesis, Columbia.

Includes bibliographical references.
1. Arithmetic--Study and teaching (Elementary)--United States. 2. Mathematical ability--Testing.
I. Title. II. Series: Columbia University. Teachers College. Contributions to education, no. 58.
QA135.5.K57 1972 372.7'3'044 70-176955
ISBN 0-404-55058-4

Reprinted by Special Arrangement with Teachers College Press, New York, New York

From the edition of 1913, New York
First AMS edition published in 1972
Manufactured in the United States

AMS PRESS, INC.
NEW YORK, N.Y. 10003

ACKNOWLEDGMENTS

No acknowledgments I can make adequately show my obligations to Dr. Edward L. Thorndike for assistance in connection with this study. To him I am indebted for suggesting and outlining the study, for guidance during its progress, for assistance in computing and interpreting the results, and for reading and revising the manuscript.

To Dr. J. McKeen Cattell and Dr. R. S. Woodworth I am indebted for valuable suggestions received during reports on the study in their seminar. To Dr. C. L. Robbins I am under obligations for careful checking of the statistical results, and to Dr. H. W. Reddick for a careful reading of the manuscript. To the children, teachers, principals, and superintendent of the Children's Aid Society of New York City, I am indebted for cheerful coöperation during the progress of the practice. To my sister, Miss Nelle Kirby, I am indebted for conducting the practice of one class reported in this study.

T. J. K.

CONTENTS

Chapter		Page
I	THE ADMINISTRATION OF THE EXPERIMENTS	1
	Children Participating	1
	Conductors of the Practice	2
	Material Used	2
	Plan of the Practice	3
	Method of Conducting the Practice	6
	Method of Scoring	9
	Efforts for Uniformity	10
II	IMPROVEMENT IN THE GROUP AS A WHOLE	12
	Addition	12
	Initial Ability	15
	Accuracy	16
	Gross Gain in Number of Problems Correctly Added	18
	Relative Gain in Addition from Sixty Minutes of Practice	20
	Gross Gain in Accuracy in Addition	22
	Summary	24
	Division	24
	Initial Ability	27
	Accuracy in Division	28
	Relative Gain in Division	31
	Gross Gain in Accuracy	32
	Summary	34
	The Value of the Practice Experiment as a Method of Teaching	35
	Factors Contributing to the Improvement	37
III	THE EFFECT OF THE DISTRIBUTION AND LENGTH OF WORK PERIOD UPON THE RATE OF LEARNING	45
	Plan of Practice	45
	Addition	47
	Initial Ability of the Groups	47
	Gross Gain	49
	Percentile Gain	51
	Gain in Accuracy	52
	Summary	54
	Division	55
	Initial Ability	55
	Gross Gain	59
	Percentile Gain	60
	Gain in Accuracy	60
	Summary	60
	Improvement in Relation to Initial Ability	68
	General Summary	69

Chapter		Page
IV	THE PERMANENCE OF THE PRACTICE EFFECT	71
	Permanence of Association Normally Used in School Work	72
	Addition	72
	Gross Gain	73
	Gain in Accuracy	73
	Division	75
	Gross Gain	75
	Gain in Accuracy	76
	Summary	77
	Permanence of Associations Through Summer Vacation	78
	Addition	78
	Gross Loss	78
	Loss Per Cent	78
	Loss in Accuracy	79
	Division	79
	Gross Loss	80
	Loss Per Cent	81
	Loss in Accuracy	81
	Permanence in Terms of Advance Over the Initial Practice-Period of the Original Experiment	81
	Addition	81
	Gross Gain	81
	Change in Accuracy	82
	Division	83
	Gross Gain	83
	Loss in Accuracy	84
	Permanence of Associations in Addition and Division during Vacation as Shown by the Amount of Practice Required to Restore these Associations to their Previous Efficiency	86
	Addition	86
	Time of Practice	86
	Gross Gain	87
	Accuracy	87
	Division	89
	Time of Practice	89
	Gross Gain	90
	Accuracy	90
APPENDIX I.	The Use of the Method of the Practice Experiment in Teaching Handwriting and Spelling	93
APPENDIX II.	Sample of Addition and Division Sheets Used	97

PRACTICE IN THE CASE OF SCHOOL CHILDREN

CHAPTER I

THE ADMINISTRATION OF THE EXPERIMENTS

Children Participating

The practice on which this study is based was conducted during the years 1911 and 1912 in the schools of the Children's Aid Society in the third- and fourth-year classes as a part of the regular grade work in arithmetic. In all, about 1350 children took part. It was the intention that the study should be conducted under school conditions normal for children, teachers, and supervisor, in order to meet, as far as possible, a current criticism that results obtained from studies with small groups of persons under laboratory conditions, are not applicable in school conditions; and also in order to establish greater confidence in the validity of the results of the study and insure the applicability of its results to actual school-room problems. It is true that one meets a greater number of factors under such conditions that can not be controlled than in an experiment conducted under laboratory conditions, but they are the factors that enter into all school-room work and help to determine both its quality and quantity. There was approximately the same similarity of ability in the classes that took the practice that one ordinarily finds in a school system where a common course of study is followed and where all the teachers are under similar supervision, but where there is diversity of nationality and extreme individual variations among the children.

Two different experiments were conducted, one in addition, the other in division. The addition was given to classes in the fourth year of the elementary school; the division, to classes in the last half of the third year and the first half of the fourth year. These classes had done the work outlined in arithmetic in the New York City course of study. Hence the children had been taught the facts and processes involved in the experiment.

Their exact proficiency in these functions at the beginning of the experiment will be defined more accurately later by means of their records in the first practice-period.

Conductors of the Practice

Thirty-nine classes took part in the practice. The writer conducted the entire practice in thirty-four of these classes. In four other classes (VI, X, XXV, and XXVI) he had charge of the initial and final practice-periods and of some others, but some of the intervening practice-periods were in charge of the principal of the school or the teacher of the class, who followed the same plan as the writer. They could do this because they were present during all the previous periods of practice and had been asked to note the exact method of procedure in order that they might be able to duplicate it exactly when the writer could not be present. Their interest and fine spirit of cooperation insured the most careful observance of all the requirements of the experiment. Since the writer usually gave the practice-period preceding and following one given by them, it was possible to note any variations that might occur in results. No greater variations were noticeable in the performance of these classes at such times from one day to the next than when the writer conducted the tests continuously. In the case of class XX, the complete practice was conducted by the class-room teacher in a school outside of New York City. All the details of the experiment were carefully observed. The results in this class conformed so closely to those in other classes that it seemed perfectly justifiable to use them in order to make the number of classes in the different groups equal. Anyone who wishes to do so, will find that the results are practically uninfluenced by omitting these five classes from the totals.

Material Used

In the addition practice, the Thorndike 'Addition Sheets' (which consist of seven different sheets, each containing 48 columns of one-place numbers, each column containing ten addends, 1's and 0's omitted, so arranged that any successive five of the columns are of nearly equal difficulty) were used. By using the seven different sheets, the chance of remembering answers to any column because of its position is reduced to a minimum. These sheets are printed in large clear type on unglazed paper which obviates eye-strain with its attending fatigue, and gives the best possible conditions for rapid, concentrated, accurate work. A sample of one of these sheets may be found on page 97.

For the practice in division, sheets were devised by the writer on the "Remainder Division Table" plan worked out by Thorndike in his "Exercises in Arithmetic," Nos. 2 and 3. These division combinations include the entire series from "20 = — 3s and — remainder" up to "89 = — 9s and — remainder," thus involving not only the combinations which are the inverse of the multiplication tables through 9 times 9, but also the division of the intervening series of numbers which give a remainder in the answer. No dividends smaller than twenty were used because they involve associations which are usually well established in children's minds and so need no further drill. A sample of these sheets may be seen on page 98. Three sheets of these combinations were arranged in such a way as to make any group of eight or ten successive combinations practically equal in difficulty to any other such group. This arrangement was the only means of obviating the inequality of difficulty in individual combinations. An examination of these sheets will convince one that the difficulty involved in any continuous group of eight or ten combinations is, for all practical purposes, the same as that for any other similar group, even though the individual combinations may differ widely in degree of difficulty. The records made by classes on these tests are further evidence of the equality of different groups of eight or ten successive combinations.

Plan of the Practice

In the addition experiment each class practiced adding these columns for 75 minutes; but this time was distributed differently for different classes. Each class had an initial practice-period of fifteen minutes. These were identical in character with the intervening practice-periods; but besides serving as practice-periods, they served as measures of ability at the beginning and end of practice. From them the change in ability for each individual was measured. The intervening practice of forty-five minutes between the initial practice-period and the final practice-period was broken up for different groups of classes in four different ways. For Group I it was divided into two practice-periods of 22½ minutes each; for Group II, into three practice-periods of 15 minutes each; for Group III, into eight practice-periods, seven of six minutes each and one of three minutes; for Group IV into twenty-two practice-periods, twenty-one of two minutes each and one of three minutes. The following statement of this plan in tabular form adds to clearness:

Groups	Initial Period	Intervening 45 Minutes	Final Period
I	15 min.	2 22½ min.	15 min.
II	15 min.	3 15 min.	15 min.
III	15 min.	7 6 min. and 1 3 min.	15 min.
IV	15 min.	21 2 min. and 1 3 min.	15 min.

It is of special importance for an understanding of the entire discussion to get clearly in mind that all the classes practiced for the same aggregate amount of time. It is further to be carefully noted that for all classes the initial period and final period of practice were of equal length. Since all four groups had the same initial and final practice-periods, but had the intervening forty-five minutes of practice distributed in different ways, any difference in the results for the groups will be a consequence of distributing a given amount of time in periods of different length.[1]

In the division experiment the aggregate time of practice was sixty minutes for each class. This was divided into periods of different lengths for different groups of classes. Each class had an initial practice-period of ten minutes and a final period of ten minutes; and the intervening practice of 40 minutes was divided in three different ways. In the division, as in the addition, the initial and final practice-periods were identical in character with the intervening practice-periods; but besides serving as practice-periods, they served as measures of ability at the beginning and end of practice from which any change in ability was measured. The first group had two intervening practice-periods of twenty minutes each; a second group had four intervening practice-periods of ten minutes each; a third group had twenty intervening practice-periods of two minutes each. The following tabular statement helps to keep the plan for each group in mind:

[1] Except for a factor to be noted on page 62 ff.

Groups	Initial Period	Intervening 40 Minutes		Final Period
I	10 min.	2	20 min.	10 min.
II	10 min.	4	10 min.	10 min.
III	10 min.	20	2 min.	10 min.

The plan of having each class drill for the same aggregate length of time, sixty minutes, makes it possible to measure the gain in all classes for this amount of drill; while breaking the forty minutes intervening between the initial test and the final test into periods of different lengths gives the opportunity, as in the addition experiment, to find the comparative effect of drill-periods of different lengths, which aggregate a constant amount.

As nearly as normal school conditions would permit, all practice-periods were given one each on successive school days to any class. So in addition, in the classes of Group I, four days were required to complete the experiment; of Group II, five days; of Group III, ten days; and of Group IV, twenty-four days. In division four days were required for the classes of Group I; six days for Group II; and twenty-two days for Group III. In three classes it was found necessary to give a two-minute practice-period twice a day, one in the morning and one in the afternoon for the last five days in order to complete the experiment within a time made necessary by school conditions. No noticeable effect was produced by this change. The practice was given to any individual class as nearly at the same time of day as was possible. In general the time of day did not vary for any class more than half an hour, but sometimes it was necessary to give a practice-period to a class in the afternoon instead of in the morning. The teachers felt that children would not do so well in the afternoon, but the results showed little or no influence from such changes of time. Of course different classes did not have their practice at the same time of day. Some classes had most of their practice early in the morning, others later in the morning. Some classes had most of their practice early in the afternoon while others had theirs late in the afternoon. Some of the very best records were made late in the afternoon despite the general feeling on the part of the teachers that a class could not do as well late in the day. Some classes seemed to be affected by the loss of school work on Saturday and Sunday, or by the intervention of a holiday. Yet in other classes remarkable gains were made following such a holiday, which seems to obviate the need of carefully evaluating any such variable factors. In fact other studies have shown that where interest is keen children do as effective work late in the day as early. Hence this factor receives no consideration.

The fact that the interval between the initial and final practice-periods was larger when the intervening practice-periods were shorter and so gave more opportunity for practice out of school and for the influence of regular school training outside the specified practice of the experiment, conditions all the results of this study. There were three possibilities. The practice could be given as it was; or it could be given with longer intervals when the practice-periods were long, so as to have the final practice-period always come, say, 24 school-days after the initial practice-period; or it could be given daily as it was, but with a varying interval between the next to the last and the last practice-period so that the final practice-period would always come, say, 24 school-days after the initial practice-period. Each method has its disadvantages. The method used here gives the pupils practicing in shorter periods greater chances of help from practice outside the experiment. The second method mixes the effect of interval-length with the effect of practice-period-length. The third gives the shorter periods the advantage of less forgetting. In interpreting the results given in this monograph, the reader may allow for the greater chance of outside practice in the short-period groups as seems wise.

The Method of Conducting the Practice

The purpose of the practice was explained to the children as carefully as possible, and in the same way to all classes. The fact that much of arithmetic work both in school and out involved addition was impressed as well as the need for speed and accuracy in it. They were told that this work would help them improve their addition. The value to be derived from the practice was indicated. In the case of addition, no preliminary work was given, because it could be assumed that any fourth-year child knew how to add columns of figures. Before the papers were passed out the children were shown one of the sheets and instructed to place their answers on the sheet which would be given to them, in the space provided. They were told to add as rapidly as possible, but to make as few mistakes as possible, as all mistakes would count against their scores. The need of accuracy combined with speed was impressed further by showing them the method of scoring, no credit save for each column correctly added. When asked what they were to do, their most common response was, "Add as fast as you can and don't make mistakes." They were also told how many minutes they were to work, but they had no means of knowing the time, once the signal to begin was given, until the signal to stop was given.

The papers were passed out by the class teacher in her regular way, with the printed side down. In all the practice each child was supplied with more problems than he could finish in the allotted time. The children placed their names, age, and grades on the back of the sheets. Then they were instructed again to add as rapidly and as accurately as possible and to place their answers on the sheets beneath the columns. They were told that the signal for beginning would be "Go," and for ending "Pencils down." The oft-repeated injunction, "Look only at your own paper" was given, and was followed quite conscientiously. After a practice-period was finished and pencils were laid down, the children were asked about the number of problems they had worked. This was done because the children were eager to tell their own scores and to learn the scores of others. It acted too as an immediate reward and so as an incentive. Care was taken to avoid undue excitement, but an attempt was made to inspire the children to their best individual efforts.

Just before beginning the second and each following practice-period, the exact score of each child in the preceding day's practice was read, both the number of columns worked and the number correct.[2] High scores with few errors were commended. If some score showed that perhaps this child was adding beyond his best speed, as indicated by the number of errors, he was advised, not in a fault-finding way but for the good of his own record, to try for greater accuracy with the hope of improving his score. These facts were put in concrete form for the children as follows: "Which gives the better score, 20 with 4 wrong or 16 with 1 wrong?" Of course this warning was not

---

[2] With a very few exceptions.

always pertinent, though usually a large number of mistakes indicated that the child was working beyond his own norm of speed. At all times during the practice the value of accuracy was stressed; and the undesirability of working an excessive number of problems with many errors was pointed out.

Further instructions calculated to maintain effort were given just before beginning the second practice-period and repeated always at the beginning of each subsequent one. The children were told that their individual improvement was to be measured and they were shown that no matter how low or high their present record their final standing would be determined by the amount of gain made. They were shown that it was not primarily a contest among the individuals of the class, but an effort on the part of each one to surpass his own previous record. The children were encouraged to compare their last record with their own previous records, and at times the scores were read to them in such a way as to indicate gains made. A hypothetical curve was drawn upon the board to indicate the ascent that would result from supposed gains, as well as to show them how to keep their own individual curves.

After a few periods of practice, actual scores were used on the board in the construction of these curves. No doubt, had each child been encouraged to draw his own curve from day to day, this concrete showing of his gain would have resulted in added effort that would have justified fully the time spent. The desire to surpass one's own previous record was the incentive most appealed to; still the spirit of rivalry was strong. One little girl expressed it well for the class by saying: "Why, you just try to beat yourself," but an ambitious boy, when asked what record he was trying to surpass, pointed to his neighbor behind him and said: "I'm trying to beat him." The average number of problems done correctly in the preceding practice-period by boys and girls as separate groups was also read to the class before beginning the day's practice. This group comparison served as an additional incentive, but would not satisfy the children as a substitute for their individual scores.

In division, preliminary work was given before the initial practice-period was begun. The need of knowing the multiplication and division combinations, both for school and life, was explained. The children were told that this work would aid them in both. Then many of the combinations were placed upon the blackboard to insure familiarity with the form of expression used in the sheets. Particular attention was given to assisting any child who appeared not to understand. After a seemingly general response was obtained in this way, in order to make doubly sure that each individual knew exactly what he was to do and how to record the result properly on the sheet, one-sixth of a sheet was given to each child on which to fill in the results. During this part of the preliminary trial very great care was taken by the writer and the grade teacher to see that all the children understood fully what was to be done. Despite this care the first papers of some few individuals could not be used because they showed clearly a failure to grasp the situation. In most cases these failures were from children who had recently entered the classes and were not yet properly classified, or who had not sufficient mastery of English to understand the instructions. Yet some failed because they were backward in learning the process of short division. Almost without exception the children who failed to grasp the situation after the explanation mentioned above, were found to be unable to handle short division in any form. Later the teacher explained the matter further to these children. They continued in the practice, but their records did not enter in the final computation. In all other respects, the same plan was followed in the division as has been described in the addition, though greater stress could be placed on speed, inasmuch as the percentage of accuracy showed a gradual increase along with the absolute gain in number of combinations.

Method of Scoring

In scoring the addition, credit was given according to columns. The score for any paper was found by deducting one from the total number of columns added for each column incorrectly added. If a child added 16 columns with 2 wrong, his score was taken as 14. Thus efficiency was measured by the rapidity and accuracy of getting the sums. In division a combination was taken as a unit, correct quotient and remainder being required for any credit. The score for any paper was found by deducting one from the total number of combinations solved for each one incorrectly solved. If a child did 30 combinations with 2 incorrect, his score was 28. One might contend that some credit should be given for a correct quotient with an incorrect remainder; but after all in the use of such combinations the result as a whole determines efficiency. Thus ability was measured by the rapidity and accuracy of getting the quotients and remainders. These methods of scoring whereby speed and accuracy are reduced to a single variable, are entirely arbitrary, but the author feels reasonably certain that any system of weighting errors which might be used would not materially change the results on which conclusions are based.

The record of no child could be used if he was absent at a first or last practice, inasmuch as the change in his ability could not be measured. But records of those who missed one or more intervening tests were used.[3] In all, about 1350 children were tested. The smallest number of papers scored for any child who was present every day during the time an experiment was in progress was four, and the greatest number twenty-four. This means that in all about 15,000 papers were graded and their scores recorded in the process of giving the two experiments.

Efforts for Uniformity

Special effort was made to keep the conditions uniform in all the classes. Each teacher was requested to give no drill during the experiment in the line of work being tested, since one purpose of the work was to measure as accurately as possible the amount of improvement her children could make with this special method of work in the exact time allotted for the practice. A fine spirit of cooperation prevailed which insured compliance with this request. No mention was made to the children about practicing for fear it might act as a suggestion. In a few cases children became so enthusiastic that they did some work at home. However, they had none of the sheets that were used in the experiment and so had to devise their own means for any practice carried on. This can only be taken as an indication of the interest

[3] This was perhaps an error of administration, but its influence on any result to be stated was very slight; hence, having begun in that way, I have not thought it worth while to undergo the labor of computing the results separately for those who were present at every single practice-period.

displayed and must be considered, as I have already said, as a factor that can not be completely controlled in a school experiment.

Perhaps the greatest difficulty in carrying on the tests uniformly for all classes was met in trying to give the instructions and appeals to different classes in such a way as to cause them to react with an equal amount of enthusiasm for the work. At all times every incentive was used that would seem to cause the children to surpass their previous records. When one bears in mind that the writer was the regular supervisor equally acquainted with all the classes, controlled by no preconceived notions of results, and interested in the experiment as a piece of accurate school work, it will be seen that the conditions would naturally be kept as nearly uniform in all classes as is possible in such a piece of work as has been done.

Some one may express surprise at not finding here a printed set of instructions given orally or read to all classes in the same way to insure uniformity. However, the author aimed not so much at uniformity in kind and amount of instruction as at uniformity of understanding and enthusiasm on the part of the different classes for the task in hand. While this plan introduced again the judgment of the writer to determine uniformity of understanding and enthusiasm, he feels that, even so, less substantial variability resulted in the total reactions of the classes measured, than if the same instructions had been uniformly read or given orally at all times. In making this effort for uniformity of understanding and enthusiasm, no great variations in instructions and appeals were necessary; and, so far as could be judged, much the same intensity of effort obtained in the performance of any class from day to day, and in the performance of different classes.

CHAPTER II

IMPROVEMENT IN THE GROUP AS A WHOLE

Addition

This chapter consists of a consideration of the results of the practice already described, in so far as they pertain to the group as a whole. The large problem to be presented is the amount of improvement in the functions tested, made by these school children of the third and fourth year in about one hour of drill with the practice experiment as the method of class work.

To measure the improvement in a group of persons subjected to a given amount of practice in any phase of activity, it is necessary to devise methods and materials which will exercise the function in question and to obtain exact and comparable measures of their ability at the beginning and end of the practice. The methods and material have been discussed in a previous chapter. We shall proceed to a presentation of the data which give the initial and final ability of the entire group and hence the basis for measuring any change which resulted from the practice. The clearest presentation of the results of this study would demand that all the class records be published in detail. But since these records alone would require more than half the space ordinarily allotted to such a report as this, only a sample record can be published. However, from this record the reader can obtain a thorough understanding of the exact sources from which all summaries presented later were obtained. The table is a complete record of one class in addition in which all the practice-periods were 15 minutes in length. Such a record is presented because it affords the best opportunity to see the progress from one period of practice to the next.

Table I shows that in the initial practice-period boy A added 85 columns, 81 of which (or 95 per cent) were correct. In the second period he added 105 columns, 97 of which were correct. In the third period he added 110 columns, 102 of which were

TABLE I

RECORD OF CLASS I IN FIVE FIFTEEN-MINUTE PRACTICE-PERIODS IN ADDITION

Indi-vid-uals	Initial Practice-Period			2nd Practice-Period		3rd Practice-Period		4th Practice-Period		Final Practice-Period			Gain		
Boys	S.	C.	%C.	S.	C.	S.	C.	S.	C.	S.	C.	%C.	Gross	Per cent	Ac.
A	85	81	95	105	97	110	102	115	107	118	116	98	35	43	+ 3
B	38	30	79	39	27	44	32	46	38	61	43	70	13	43	— 9
C	83	81	98	80	53	85	61	98	74	92	77	83	—4	—5	—15
D	40	27	68	39	31	48	36			57	34	42	7	26	—26
E	49	42	89	57	51	64	59	76	64	80	67	83	20	43	— 6
F	80	74	93	90	89	101	99	119	117	112	112	100	38	51	+ 7
G	51	49	96	68	67	75	71	89	86	98	97	99	48	98	+ 3
H	58	51	88	49	41	54	45	54	46	57	43	75	8	16	—13
I	48	46	98	65	49	55	54	52	50	59	57	97	11	24	— 1
J	37	26	70	44	39	43	40	48	45	59	50	83	24	92	+15
K	89	82	92	40	82	99	92	112	104	96	91	95	9	11	+ 3
L	112	109	97	128	127	138	136	140	132	156	148	95	39	36	— 2
M	42	38	90	47	42	51	45	49	42	51	47	92	9	24	+ 2
N	30	27	90	41	40	45	44	47	44	50	45	90	18	67	0
O	24	15	63	35	19	52	17	74	25	51	27	53	12	80	—10
P	50	49	98	60	57	76	75	80	77	95	91	95	42	86	— 3
Q	37	29	78	41	31	46	37	28	20	48	39	81	10	36	+ 3
R	24	17	71			40	25	39	30	37	27	75	10	59	+ 2
S	33	27	82	32	30	37	32	55	47	48	39	81	2	7	— 1
Girls															
A	19	5	74	24	4	29	6	31	5	34	15	47	10	200	—27
B	42	31	74	38	37	24	22	22	20	52	48	83	12	39	+ 9
C	46	37	80	50	38	55	44	61	52	61	48	79	11	30	— 1
D	30	18	60	38	30	34	26	37	26	37	32	87	14	78	+27
E	32	31	97	44	41	56	48	67	64	73	69	95	38	123	— 2
F	19	16	84	43	41	50	45	55	47	78	77	99	61	381	+15
G	26	20	75	44	30	33	19	49	32	36	25	69	5	25	— 6
H	45	39	87	54	47	61	57	84	79	96	85	89	46	118	+ 2
I	23	21	91	24	15	28	23	25	13	37	32	87	11	52	— 4
J	35	25	71	40	35	53	48	60	55	64	59	92	34	136	+21
K	56	52	93	71	66	54	54	68	67	63	61	97	9	17	+ 4
L	25	18	72	33	22	38	33	48	35	40	37	93	19	106	+21
M	32	26	81	45	36	49	47	64	63	70	65	90	39	150	+ 9
N	26	19	73	36	30	35	30	48	33	52	36	70	17	89	— 3
O	75	74	99	80	80	83	77	89	79	96	89	93	15	20	— 6
P	37	36	97	40	40	41	39	45	43	40	36	90	0	0	— 7
Q	37	35	95	30	26	35	34	37	30	49	44	90	9	26	— 5
R	45	37	82	56	54	59	55	62	58	71	65	92	28	76	+10
S	32	30	94	40	35	40	38	40	39	48	41	85	11	37	— 9
T	34	31	91	38	35	48	42	56	47	56	41	75	10	32	—16
1	*2*	*3*	*4*	*5*	*6*	*7*	*8*	*9*	*10*	*11*	*12*	*13*	*14*	*15*	*16*

S.=Problems solved.
C.=Problems correct.
%C.=Per cent of problems correct.
Ac.=Accuracy.

correct. In the fourth period he added 115 columns, 107 of which were correct. In the final period he added 118 columns, 116 of which (or 98 per cent) were correct. His gross gain was 35 columns. That is, he added correctly 35 more columns in the final fifteen-minute period than in the initial fifteen-minute period. This gross gain of 35 columns was on an initial ability of 81 columns; and hence means a gain per cent of 43. In the initial fifteen-minute period he add 95 per cent of his columns correctly; and in the final fifteen-minute period he added 98 per cent of his columns correctly which made a gain of 3 per cent in accuracy. Each of the columns of the table has been given a number (printed at the bottom) by which it will be referred to in the future discussion.

Referring to these small figures in italics at the bottom of the columns, the reader can gain a complete interpretation of the entire record from the following explanation, bearing in mind also that the letter " S " at the top of a column designates the entire number of problems solved, " C " the number of problems correct, " %C," the per cent of problems worked correctly, "Ac," accuracy, and that the word " Gain " at the head of the last three columns is to be taken with each. Column *1* gives the individuals, boys and girls separate. Column *2* gives the entire number of problems solved in the initial 15-minute period; column *3* gives the number of problems that were correct in the same period; and column *4* gives the per cent of problems worked correctly, or the per cent of accuracy in the initial period. The percentages of accuracy in column *4* were found by dividing the number in column *3* by the corresponding number in column *2*. Columns *5* and *6*, *7* and *8*, *9* and *10* give the entire number of problems worked and the number right in the three intervening 15-minute periods. These pairs of columns correspond to *2* and *3* in the initial practice-period. The per cent of accuracy for these intervening tests was not necessary for later discussion and so was omitted. Columns *11*, *12* and *13* of the final period correspond to columns *2*, *3*, and *4* in the initial period. Column *14* gives the gross gain in number of problems correct or the absolute gain, and is the algebraic difference between the number in column *3* and the corresponding number in column *12*. Column *15* gives the gain expressed in hundreths obtained from dividing the number in column *14* by the corresponding number in column *3*. Column *16* is the gross gain in accuracy expressed in per cent of problems correct and is the algebraic difference between the number in column *4* and the corresponding number in column *13*. It is to be noted that this last column is not a gain per cent as is column *15*, but the gross gain in accuracy expressed in a per cent. A minus sign indicates a *loss* wherever used.

Every other class record in addition used in this study corresponds closely to this one in the categories of initial record, final record, per cent of accuracy, gross gain, percentile gain, and per cent gained in accuracy. The only difference in other class records results from the fact that the intervening periods of practice were of different lengths, some being 22½ minutes, some 6 minutes, and some 2 minutes in length.

For the entire addition experiment, there are 21 class records. These complete records are placed on file in Teachers College where they are accessible to any one wishing to use them.

Initial Ability

The first group of facts to be considered in arriving at a solution to the problem with which this chapter deals, provides data which give a measure of the ability of the entire group in addition at the beginning of practice. All the individual records of "problems added correctly" in the initial period of which those in column *3* of Table I are samples, have been distributed, class by class, boys and girls separate. A summary of these facts affords the data necessary to show the distribution, central tendency, and variability of the group as a whole, as well as the reliability of the measures used.

Table II gives a distribution of the number of problems worked correctly in the initial fifteen-minute period by 732 fourth-year children. The number of problems worked is given in groups of 5. The table shows that 3 children (or 0.4 per cent of the entire group) worked from 0 to 4 problems correctly in fifteen minutes; 50 children (or 6.8 per cent) worked from 5 to 9 problems correctly; 80 children (or 10.9 per cent) did from 10 to 14 problems, etc. The figures show that the distribution is decidedly skewed toward the high end. The table shows the range in number of problems added correctly by these fourth-year children to be from '0 to 4' problems up to '105 to 109' problems. The median number of problems worked correctly is 23.3; that is, there are as many children who worked 23.3 problems or more as there are who worked 23.3 problems or less. The variability of the group is further shown by the P. E., 7.8, a

TABLE II

NUMBER OF PROBLEMS ADDED CORRECTLY IN THE INITIAL FIFTEEN-MINUTE PERIOD

Problems added correctly....	0 to 4	5 to 9	10 to 14	15 to 19	20 to 24	25 to 29	30 to 34	35 to 39	40 to 44	45 to 49	50 to 54	55 to 59
Number of individuals......	3	50	80	133	130	111	81	57	33	14	13	7
Per cent.......	.4	6.8	10.9	18.1	18	15.2	11	7.8	4.5	1.9	1.8	.9

TABLE II—*Continued*

Problems added correctly....	60 to 64	65 to 69	70 to 74	75 to 79	80 to 84	85 to 89	90 to 94	95 to 99	100 to 104	105 to 109	Total
Number of individuals......	4	3	7	0	3	1	1	0	0	1	732
Per cent.......	.5	.4	.9	0	.4	.1	.1	0	0	.1	100

Median	23.3	problems
25 Percentile	16.5	
75 Percentile	32.1	
P.E.	7.8	
P.E. $_{\text{t.-obt. Av.}}$	.29	

number which when subtracted from and added to the median, gives the approximate limits within which the middle fifty per cent of the class falls. In other words, when the lowest twenty-five per cent and the highest twenty-five per cent of the class are cut out, the limits within which the remaining middle fifty per cent falls, are 16.5 columns and 32.1 columns. The per cents in the table show that 81 per cent of the children ranged from 10 to 39 problems; 51 per cent from 15 to 29 problems.

Accuracy

In order to define more exactly the initial ability of this group of children in addition it is necessary to know the degree of accuracy of their performance. The accuracy for each child was determined by dividing the number of problems worked correctly by the entire number worked. Just what these per cents of accuracy are the reader may see from Table I, column *4*.

The following table is a distribution of the per cents of accuracy of the entire group of 732 children, of which those per cents in column *4* of Table I form a part.

TABLE III

PER CENT OF PROBLEMS ADDED CORRECTLY IN THE INITIAL FIFTEEN-MINUTE PERIOD

Per cent of accuracy...	6 to 10	11 to 15	16 to 20	21 to 25	26 to 30	31 to 35	36 to 40	41 to 45	46 to 50	51 to 55
Individuals...........	1	1	0	2	3	9	13	16	22	17
Per cent of group......	.1	.1	0	.3	.4	1.2	1.8	2.2	3	2.3

TABLE III—*Continued*

Per cent of accuracy...	56 to 60	61 to 65	66 to 70	71 to 75	76 to 80	81 to 85	86 to 90	91 to 95	96 to 100	Total
Individuals...........	25	46	58	84	100	81	108	76	70	732
Per cent of group......	3.4	6.3	8	11.5	13.7	11.1	14.8	10.4	9.6	100%

Median	79.	per cent
25 Percentile	68.	
75 Percentile	89.	
P.E.	10.	
P.E. $_{\text{t.-obt. Av.}}$	.38	

Table III gives the per cents of accuracy in groups of five. It shows that 70 children (or 9.6 per cent of the entire group) added correctly from 96 to 100 per cent of the entire number of problems which they worked; 76 added correctly from 91 to 95 per cent of the problems which they worked, etc. These figures show that the distribution is skewed toward the low end. The range in per cent of problems added correctly is from 6 to 100. The median per cent of accuracy is 79, which shows that there were as many children who added correctly four-fifths or more of their problems as there were who added correctly four-fifths or less. The upper 25 per cent of the class added with an accuracy of 89 per cent or more; the upper 60 per cent of the class added with an accuracy of 68 per cent or less. The range of accuracy for the middle 50 per cent of the class was from 68 per cent to 89 per cent.

By combining now the data derived from Tables II and III, the initial ability of the group can be defined both in terms of amount and accuracy of performance. We can give the clearest idea of the initial performance of this group in addition by saying that the median ability in number of problems worked correctly in fifteen minutes was 23.3 problems, which was 79 per cent of the entire number of problems worked. Or, taking as a basis the entire number of problems worked instead of those worked correctly, one can say that the chances are that one child in any two picked at random would be able to add 29.5 columns or more in 15 minutes with an accuracy of 79 per cent or more.

Gross Gain in Number of Problems Correctly Added

We are now ready to consider the gain made in problems correctly added by this group of fourth-year children in the course of seventy-five minutes of drill with the practice experiment as the method of class work.[4] This gain will be considered in two ways, absolute gain and percentile gain. The present consideration has to do with the absolute gain. It will be recalled that the initial test and the final test for all classes were of equal length, 15 minutes, and that between the initial and final tests all classes practiced for the same length of time, 45 minutes. It is now our problem to measure the amount of improvement that the entire group showed in the final 15 minutes of practice over the initial 15 minutes.

Referring again to Table I, columns *3, 12* and *14,* the reader will understand clearly the exact measures that are involved in this part of the discussion. It will be seen that boy A in the initial test had 81 problems correct; in the final test he had 116 problems correct which gave him an absolute gain of 35 problems in the final 15 minutes over the number of problems worked correctly in the initial 15 minutes. So, boy B gained 13 problems, girl A 10 problems, etc. All of these gross gains were

[4] The gain is measured for only 60 minutes of practice. The initial period is 15 minutes and the final period is 15 minutes. The record for each of these periods gives the adding rate at the middle of each period. Hence the time of practice that is measured is from the middle of the initial period to the middle of the final period, or 60 minutes.

distributed in a table, class by class, boys and girls separate. Only a summary of this distribution for the entire group is presented here.

TABLE IV

GROSS GAIN, IN NUMBER OF COLUMNS ADDED CORRECTLY IN FIFTEEN MINUTES, RESULTING FROM SIXTY MINUTES OF PRACTICE

Columns gained...	—15 to —12	—11 to —8	—7 to —4	—3 to 0	1 to 4	5 to 8	9 to 12	13 to 16	17 to 20	21 to 24	25 to 28
Individuals.......	5	11	28	63	93	105	111	90	75	29	29
Per cents........	.7	1.5	3.8	8.6	12.7	14.3	15.2	12.3	10.2	4	4

TABLE IV—*Continued*

Columns gained...	29 to 32	33 to 36	37 to 40	41 to 44	45 to 48	49 to 52	53 to 56	57 to 60	61 to 64	65 to 68	Total
Individuals.......	25	24	16	5	7	5	2	3	4	2	732
Per cents........	3.4	3.3	2.2	.7	.9	.7	.3	.4	.5	.3	

Median	10.7	columns
25 Percentile	3.6	
75 Percentile	18.8	
P.E.	7.6	
P.E. t.-obt. Av.	.28	

Table IV gives a summary of the distribution of the gross gain in number of columns added correctly in fifteen minutes resulting from 60 minutes of practice. The number of problems gained is given in groups of 4. The table reads: "Five children (or 0.7 per cent of the entire group) gained from —15 to —12 problems." This means, of course, that they lost. The other end of the table reads: "Two children (or 0.3 per cent of the entire group) gained from 65 to 68 problems; four children (or 0.5 per cent of the group) gained from 61 to 64 problems, etc." The figures show that the distribution is skewed toward the high end. The range for the group is from 15 to 12 problems lost to 65 to 68 problems gained. The median gain is 10.7 columns, which means that the median increase in ability was such that 10.7 more columns were added correctly in 15 minutes at the end of practice than in 15 minutes at the beginning of practice.

It also means there were just as many children who gained 10.7 columns or more, as there were who gained 10.7 columns or less. The members of the upper 25 per cent of the group gained 18.8 columns or more while the members of the lower 25 per cent of the group gained 3.6 columns or less. This leaves the range for the middle fifty per cent of the class from 3.6 columns to 18.8 columns gained. The percentages in the table show that 65 per cent of the entire group gained from 1 to 20 problems and that a larger per cent gained from 9 to 12 columns, than any other group of four columns.

Relative Gain in Addition From Sixty Minutes of Practice

Gain may be expressed in absolute terms, as in the last section; or it may be expressed in the per cent which the gross gain is of the original quantity on which it was gained. We are so accustomed to think of gain or improvement in relative terms that such terms often become a more satisfying measure of the change in question than the absolute terms. We found in the last section that the median gain in number of problems added correctly in 15 minutes was 10.7 columns. It is now the problem to find what per cent of the initial ability this gross improvement was.

Referring again to Table I, column *15*, the reader will clearly understand the source from which the present data are obtained. Boy A made a gross gain of 35 problems; that is, he worked correctly 35 more problems in the last fifteen minutes than in the initial fifteen minutes. Or he made a gain of 35 problems on an initial ability of 81 problems, or 43 per cent. So boy B gained 43 per cent, girl A gained 200 per cent and girl B, 39 per cent, etc. In almost every case these per cents are of a different initial ability, and to have their proper significance they need to be considered in relation both to the initial ability and the gross gain of the individual whose performance they represent. That is, the same gain per cent for two individuals does not necessarily indicate that their gain was equally meritorious. Taking the cases of boys A, B and E, who have gained the same relative amount, we would be justified in saying that A's 43 per cent gain represents a much higher efficiency of performance that that of B or E. Likewise we feel certain that L's 36 per cent of gain is far more noteworthy than Q's 36 per cent, and it is plainly evident that G's 98 per cent does not mean almost twice as meritorious a gain as that of F. We are badly in need of a more equitable means of measuring relative gain than is afforded by per cents, but until such a means is perfected we shall have to continue using the one now in vogue. Such considerations, however, have to do especially with individual differences which will be considered in a later portion of this study.

All the per cents which correspond to those of column *15*, Table I, were distributed class by class, boys and girls separate, but only a summary of this table is presented here, from which an adequate notion of the central tendency and variability of the group in relative gain can be explained.

TABLE V

GAIN PER CENT IN COLUMNS ADDED CORRECTLY IN FIFTEEN MINUTES RESULTING FROM SIXTY MINUTES OF PRACTICE

Gain, per cent...	-89 to -75	-74 to -60	-59 to -45	-44 to -30	-29 to -15	-14 to 0	1 to 15	16 to 30	31 to 45	46 to 60	61 to 75	76 to 90	91 to 105	106 to 120
Individuals.....	3	1	6	11	20	61	66	97	85	92	80	54	40	29
Per cents.......	.4	.1	.8	1.5	2.7	8.3	.9	13.2	11.6	12.6	10.9	7.4	5.5	4

TABLE V—*Continued*

Gain, per cent...	121 to 135	136 to 150	151 to 165	166 to 180	181 to 195	196 to 210	211 to 225	226 to 240	241 to 255	256 to 270	271 to 285	286 to 300	301 to	Total
Individuals.....	14	21	9	7	2	7	4	8	1	0	2	2	10	732
Per cents.......	1.9	2.9	1.2	.9	.3	.9	.5	1.1	.1	0	.3	.3	1.4	100

Median	48 per cent
25 Percentile	18
75 Percentile	83
P.E.	32.5

In Table V the relative gains are distributed in groups of 15. The range is from a loss of 89 per cent to a gain of 300 per cent or more. The mode is in the group 61 to 75 per cent. The median gain per cent is 48, which means that there were as many children who gained more than 48 per cent as there were who gained less than 48 per cent. The upper 25 per cent of the group gained 83 per cent or more while the lower 25 per cent gained 18 per cent or less, which leaves the range for the middle 50 per cent of the group from a gain per cent of 18 to 83. The percentages in the table show that 14 per cent of the entire group lost as a result of the practice, and that 56 per cent of the entire group gained from 16 to 90 per cent. Again bearing in mind the wide variability of the class, we can say that the central tendency of the relative gain of the group is a gain of 48 per cent.

Gross Gain in Accuracy in Addition

To determine the complete effect of the practice upon this group one other factor remains to be considered, *viz.*, the influence of the practice upon the accuracy of making the associations involved. It will be recalled from the instructions given to the classes just before each practice-period that two factors were emphasized at all times during the practice as constituent elements of the efficiency of the performance, speed and accuracy. We have seen that the group as a whole worked with enough increase of speed to add correctly 10.7 more columns in the final fifteen-minute period than in the initial period, which meant an increase in speed of about fifty per cent. The question now arises, was this increase in speed accompanied by an increase in accuracy, or was it at the expense of accuracy, or did the group as a whole add with as good a chance of getting a correct answer at the end of the practice as at the beginning?

For an understanding of the data on which the answer to this question is based, the reader is referred again to Table I, columns *4*, *13*, and *16*. The degree of accuracy of boy A at the beginning of practice was 95 per cent; at the end of practice his percentage of accuracy was 98 per cent. Hence his gross gain in accuracy expressed in percentage of answers that were correct was 3. It is to be noted that this is a gross gain and not a gain per cent. So boy B lost 9 per cent, girl A lost 27 per cent, and girl B gained 9 per cent. All these per cents were distributed class by class, boys and girls separate. A summary of this table will suffice to give sufficient data to determine the change in the group as a whole, in control tendency and variability.

TABLE VI

GROSS GAIN IN ACCURACY IN ADDITION EXPRESSED IN PER CENT OF ANSWERS THAT WERE CORRECT RESULTING FROM THE SIXTY MINUTES OF PRACTICE

Per cent gained.	–55 to –51	–50 to –46	–45 to –41	–40 to –36	–35 to –31	–30 to –26	–25 to –21	–20 to –16	–15 to –11	–10 to –6	–5 to –1	0 to 4	5 to 9
Individuals.....	8	2	2	6	19	23	17	54	63	68	101	111	79
Per cents.......	1.1	.3	.3	.8	2.6	3.1	2.3	7.4	8.6	9.3	13.8	15.2	10.8

TABLE VI—*Continued*

Per cent gained.	10 to 14	15 to 19	20 to 24	25 to 29	30 to 34	35 to 39	40 to 44	45 to 49	50 to 54	55 to 59	60 to 64	65 to 69	Total
Individuals.....	61	47	28	17	6	8	2	5	1	3	0	1	732
Per cents.......	8.3	6.4	3.8	2.3	.8	1.1	.3	.7	.1	.4	0	.1	

Median	–.4	per cent
25 Percentile	–12.4	
75 Percentile	9.2	
P.E.	10.8	

In Table VI, the gross gains in accuracy expressed in per cents are given in groups of 5. Negative signs indicate loss. The range in the change in accuracy effected by the practice is from a loss of 55 per cent to a gain of 69 per cent. The table reads, "Eight children (or 1.1 per cent) of the entire group lost from 55 to 51 per cent, etc." The figures indicate a curve which approaches close to the normal curve of distribution. The mode is at 0 to 4 per cent gain. The median is .4 per cent loss, which shows that there were as many children who lost .4 per cent or more in accuracy as there were who lost .4 per cent or less or gained in accuracy. The figures following "Individuals" show that 363 of the children lost in accuracy while 369 children either maintained their initial proficiency for accuracy or worked with greater chance of getting a correct answer. The upper 25 per cent of the group gained 9 per cent or more in accuracy while the lower 25 per cent lost 12 per cent or more, which gives the range for the middle 50 per cent of the group from a loss of 12 per cent to a gain of 9 per cent.

Summary

The initial performance in addition of the group of 732 fourth-year children showed a median ability on their part to add correctly in fifteen minutes 23.3 columns such as were used in the practice. Their median accuracy in this initial performance was found to be about 80 per cent, which shows that the median number of problems added regardless of the answer obtained was about 30 columns. From these two central tendencies we can say that the median initial ability of this group of 732 fourth-year children in addition was such that they could add correctly 23.3 columns, such as were used in the test, with an accuracy of 80 per cent.

The change effected by the practice of 60 minutes in the group as a whole was an improvement in speed such that they could add 10.7 more columns correctly in the final 15 minutes of the practice than in the initial 15 minutes, while the change in accuracy was so small as not to need consideration.

Division

The experiment in division, with children in the second half of the third year and the first half of the fourth year, was conducted on the same general plan as the one in addition with fourth-year children, except that the initial and final periods of practice were 10 minutes each instead of 15, and that the entire time spent in the practice was 60 minutes instead of 75. The intervening 40 minutes between the initial and final practice-periods were divided for different groups of classes into periods of 20 minutes, 10 minutes and 2 minutes. This plan of time required 4 school days to complete the experiment when the intervening time was divided into 20-minute periods, 6 consecutive school days when the intervening time was divided into 10-minute periods, and 22 consecutive school days when the intervening 40 minutes were divided into periods of 2 minutes each. The method of conducting the practice was described in Chapter I, and a sample sheet of the material used may be found on page 98. It is necessary to remember that in scoring these division combinations a credit of one was given for each combination if the quotient and remainder were both correct. If either was wrong, no credit was given. In other words, the score was found by deducting one from the entire number of combinations worked for each one that was incorrectly worked.

In Table VII, an exact class record is given of a class for which the intervening 40 minutes were divided into periods of ten minutes each. This arrangement made all the practice-periods of this class of equal length. Such a record is presented because it gives samples of all data to be used in this discussion and affords the best opportunity to note the change in ability from day to day. In all, eighteen classes took part in the division experiment, which means that there were eighteen such records as the one here presented, from which the data to be presented in the following discussion were obtained. Six of these 18 records involve exactly the same number of practice-periods as the one given in Table VII. Six others have but two practice-periods between the initial and the final periods and so are more brief. The other 6 have 20 practice-periods between the initial and final periods and hence occupy more than three times the space occupied by the one given in Table VII. Six hundred six third- and fourth-year children took part in the experiment, in the course of which about 6500 papers were scored and entered in these 18 records. These can not be printed here, but they are placed on record in Teachers College where any one may use them. The following table gives sufficient data to enable the reader to understand the exact sources from which the summaries that are to be presented later were obtained.

Table VII reads as follows, "Boy B in the initial practice-period worked 70 combinations, 63 of which were correct, or 90 per cent of them. In the second practice-period he worked 72 combinations with 72 correct. In the third practice he worked 80 combinations with 80 correct. In the fourth practice he worked 87 combinations with 83 correct. In the fifth practice-period he worked 99 combinations with 92 correct. In the final practice-period he worked 119 combinations with 112 correct, or 94 per cent of them. His gain in number of combinations worked correctly was 49, his gain per cent 78, and his gain in accuracy expressed in per cent was 4 per cent." So for any other individual. In the following discussion the quantities mentioned above will be referred to by the numbers in italics at the bottom of the columns. The method of finding the per cent of accuracy, the gain per cent in number of problems

TABLE VII

Record of Class VII in Six Ten-minute Practice-periods in Division

Indi-vid-uals	Initial Practice-Period			2nd Practice-Period		3rd Practice-Period		4th Practice-Period		5th Practice-Period		Final Practice-Period			Gain		
Boys	S.	C.	%C.	S.	C.	S.	C.	S.	C.	S.	C.	S.	C.	%C.	Gross	Per cent	Ac.
A	55	54	98	60	58	84	82	77	75	76	75	74	71	96	17	30	— 2
B	70	63	90	72	72	80	80	87	83	99	92	119	112	94	49	78	+ 4
C	27	25	93	45	38	55	51	59	56	66	60	70	66	94	41	168	+ 1
D	16	12	75	14	14	25	24	25	24			18	17	95	5	63	+20
E	77	77	100	99	98	106	106	112	112	133	133	153	151	99	74	96	— 1
F	27	21	78	28	18	19	13	30	23	36	32	45	42	93	21	100	+15
G	31	25	81	36	31	46	43	41	35	38	31	49	42	86	17	68	+ 5
H	50	49	98	51	48	61	61	56	53	55	53	65	63	97	14	29	— 1
I	62	62	100	62	62	69	69	69	67	60	60	68	68	100	6	10	0
J	34	30	88	48	44	58	57	54	52	52	50	53	52	98	22	73	+10
K	18	17	95	36	33	54	50	49	44	77	70	63	59	94	42	40	— 1
L	30	19	63	35	25	46	33	64	58	40	36	37	35	95	16	84	+32
M	90	86	96					104	98	114	113	128	127	99	41	48	+ 3
N	69	68	99	62	59	83	81	78	75	77	77	92	89	98	21	31	— 1
O	43	38	89	53	45	69	65	61	58	56	53	63	60	95	22	58	+ 6
P	27	27	100	27	27	33	32	31	27	28	28	30	27	90	0	0	—10
Q	78	74	95	89	86	109	99	104	104	108	98	107	98	91	24	32	— 4
R	37	37	100	58	58	67	67	62	60	69	67	77	77	100	40	105	0
S	48	44	92			63	57	55	52	66	66	77	77	100	33	75	+ 8
T	67	66	99	88	86	103	101	102	101	99	95	112	111	99	45	68	0
U	54	51	94	57	54	92	68	69	67	62	62	64	64	100	13	25	+ 6
V	35	32	92			40	37	35	32	44	41	56	54	97	22	69	+ 5
W	27	25	92	27	25	43	39	38	35	36	36	39	39	100	14	56	+ 8
Girls																	
A	14	11	79	29	21	37	29	31	19			38	31	82	20	55	+ 3
B	24	23	96	37	36	30	29	35	32	45	45	50	50	100	27	117	+ 4
C	39	36	93	39	38	52	51	53	49	50	48	57	53	93	17	47	0
D	50	46	92	59	56	69	68	62	62	51	51	81	80	99	34	74	+ 7
E	56	54	97	61	58	61	61	65	65	55	53	66	62	94	8	15	— 3
F	52	49	94	54	52	71	71	66	63	50	48	76	73	96	24	49	+ 2
G	46	41	89	49	46	51	49	50	47	49	46	56	61	91	10	24	+ 2
H	42	39	93	53	51			68	68	65	64	79	77	98	38	97	+ 5
I	76	74	98	20	19	83	83	78	77	81	79	110	107	97	33	45	— 1
J	47	46	98	50	48	69	68	55	55	55	55	66	64	97	18	37	— 1
K	10	8	81	7	2	25	24	14	7	28	26	20	13	65	5	63	—23
L	72	69	96	71	69	93	92	93	93	90	87	98	96	98	27	39	+ 2
M	45	45	100	55	54	66	62	77	76	63	54	71	67	95	22	49	— 5
N	60	57	95	59	59	72	72	57	56	70	70	84	83	99	26	46	+ 4
O	24	16	67	37	22	36	27	45	36	41	32	52	45	87	29	181	+20
P	45	45	100	39	38	59	59	64	58	62	61	71	70	99	25	56	— 1
Q	24	20	83	16	15	40	39	32	30	34	31	29	26	90	6	30	+ 7
R	42	40	95	43	42	50	50	50	49	45	44	52	49	94	9	23	— 1
S	23	20	87			29	29	36	25	33	32	37	36	98	16	80	+ 1
T	26	23	89	44	41	56	53	51	50	43	42	56	56	100	33	143	+11
U	30	16	46	30	16	37	29	32	23	37	32	38	36	95	30	125	+49
1	*2*	*3*	*4*	*5*	*6*	*7*	*8*	*9*	*10*	*11*	*12*	*13*	*14*	*15*	*16*	*17*	*18*

S.=Problems solved.
C.=Problems correct.
%C.=Per cent of problems correct.
Ac.=Accuracy.

worked correctly, and the gain in accuracy expressed in per cent is the same as that used in addition, Table I, to which the reader may refer for the method if the meaning of these figures is not clear. " S " indicates problems solved, " C " problems solved correctly, and "Ac." accuracy.

Initial Ability

To determine the initial ability of this group of 606 children of last half of third and first half of fourth year in giving the results for the division combinations, two factors must be considered,—first the number of such combinations they worked correctly, and second the accuracy of their performance. The data to determine the first factor are the number of combinations worked correctly in the initial ten-minute period, a sample of which is given in Table VII, column 3. The numbers of combinations answered correctly by the 606 children were distributed class by class, boys and girls separate, but only the following summary of this distribution table can be presented here. (Table VIII.)

TABLE VIII

NUMBER OF DIVISION COMBINATIONS ANSWERED CORRECTLY IN THE INITIAL TEN-MINUTE PERIOD

Number of combinations	0 to 4	5 to 9	10 to 14	15 to 19	20 to 24	25 to 29	30 to 34	35 to 39	40 to 44	45 to 49	50 to 54	55 to 59	60 to 64	65 to 69
Individuals	0	23	33	65	56	52	51	57	45	44	38	42	28	15
Per cent		3.8	5.4	10.7	9.2	8.6	8.4	9.4	7.4	7.3	6.3	6.9	4.6	2.5

TABLE VIII—*Continued*

Number of combinations	70 to 74	75 to 79	80 to 84	85 to 89	90 to 94	95 to 99	100 to 104	105 to 109	110 to 114	115 to 119	120 to 124	125 to 129	Total
Individuals	13	16	3	9	5	5	0	3	1	0	1	1	606
Per cent	2.1	2.6	.5	1.5	.8	.8	0	.5	.2	0	.2	.2	100

Median	34.5	combinations
25 Percentile	21.7	
75 Percentile	52.5	
P.E.	15.4	
P.E. $_{\text{t.-obt. Av.}}$	.63	

Table VIII shows a distribution (in groups of 5) of the number of combinations done correctly in the initial 10 minutes. It shows that 23 children (or 3.8 per cent of the entire group) did from 5 to 9 combinations, 33 children (or 5.4 per cent of the entire group) did from 10 to 14 combinations, etc. The range in number of combinations done correctly is from 5 to 126. The median number is 34.5 combinations. This means that there were just as many children who worked 34.5 combinations or more correctly in 10 minutes as there were who worked 34.5 combinations or less. The upper 25 per cent of the group did 53 combinations or more while the lower 25 per cent did 22 combinations or less. The percentages in the table show that 74 per cent of the group did from 15 to 59 combinations. The variability of the group is also shown by the P. E., 15.4. The data presented here give 34.5 combinations as the most likely true median.

Accuracy in Division

With what degree of accuracy did this group work the combinations which they attempted in the initial 10 minutes, is the next question for consideration. The source of the data used in answering this question can best be seen in Table VII, column *4*. The per cents of accuracy of which those in column *4* are part were distributed class by class, boys and girls separate. The following table (Table IX) shows a summary from this larger table.

Table IX reads as follows from the right side: "Two hundred and thirty-seven children (or 39.1 per cent of the entire group) worked 96 to 100 per cent of their combinations correctly, etc." The per cents are given in groups of 5. The range is from 26 per cent of accuracy to 100 per cent. The figures show almost a right-angle distribution. However, had the per cents been scaled more finely at the upper end, the distribution would then have shown a very decided skewness toward the lower end. The median per cent of accuracy is 93, that is, just as many children worked with an accuracy of 93 to 100 per cent as with an accuracy of 26 to 93 per cent. The upper 25 per cent of the children worked with an accuracy of 97 per cent or more. The lower 25 per cent worked with an accuracy of 85 per cent or less.

TABLE IX

THE PER CENT OF CORRECT ANSWERS TO THE DIVISION COMBINATIONS IN THE INITIAL TEN-MINUTE PERIODS

Per cent of correct answers	26 to 30	31 to 35	36 to 40	41 to 45	46 to 50	51 to 55	56 to 60	61 to 65
Individuals	1	0	1	3	7	5	7	12
Per cent	.2	0	.2	.5	1.2	.8	1.2	2

TABLE IX—*Continued*

Per cent of correct answers	66 to 70	71 to 75	76 to 80	81 to 85	86 to 90	91 to 95	96 to 100	Total
Individuals	11	22	31	53	70	146	237	606
Per cent	1.8	3.6	5.1	8.7	11.6	24.1	39.1	100

Median	93	per cent
25 Percentile	85	
75 Percentile	97	
P.E.	6	

The upper 63 per cent of these children worked with an accuracy of 91 per cent or more.

With the accuracy of the group determined, we can now define the initial ability of the group more exactly by saying that it was such that the median number of columns worked correctly by the group as a whole was 34.5 combinations, and according to the median per cent of accuracy 34.5 combinations were 93 per cent of the median number of combinations attempted. The chances are that one child out of two in the group taken at random would do 34.5 combinations correctly in 10 minutes with an accuracy of 93 per cent.

Gross Gain in Number of Combinations Worked Correctly

What gain in ability did this group of 606 third- and fourth-year children make in the course of 60 minutes of practice, is the next question to be answered.[5] This gain has been measured

[5] While there were 60 minutes of practice the gain was measured for only 50 minutes. The initial practice-period and the final practice-period were 10 minutes each. But the record for each of these periods gives the adding rate at the middle of each period. Hence the amount of practice whose effect is measured is from the middle of the initial period to the middle of the final period, or 50 minutes.

both absolutely and relatively. Our present consideration is the gross gain, which was found for each individual by subtracting the number of combinations worked correctly in the first 10 minutes of practice from the number worked correctly in the final 10 minutes of practice.

Referring to Table VII, column *16*, the reader will see that boy A gained 17 combinations, boy B, 49 combinations, etc. The gains made by the 606 children were distributed class by class, boys and girls separate, but for our discussion a summary of this large distribution table must suffice. This summary appears in Table X.

TABLE X

THE GROSS GAIN IN NUMBER OF COMBINATIONS ANSWERED CORRECTLY, FROM FIFTY MINUTES OF PRACTICE

Combinations gained	−19 to −15	−14 to −10	−9 to −5	−4 to 0	1 to 5	6 to 10	11 to 15	16 to 20	21 to 25	26 to 30	31 to 35	36 to 40	41 to 45
Individuals	2	5	3	15	20	49	55	60	69	60	53	41	39
Per cents	.3	.8	.5	2.5	3.3	8.1	9.1	9.9	11.4	9.9	8.7	6.8	6.4

TABLE X—*Continued*

Combinations gained	46 to 50	51 to 55	56 to 60	61 to 65	66 to 70	71 to 75	76 to 80	81 to 85	86 to 90	91 to 95	96 to 100	Total
Individuals	31	18	25	17	16	10	5	5	4	2	2	606
Per cents	5.1	3	4.1	2.8	2.6	1.7	.8	.8	.7	.3	.3	100

Median	27.5 combinations
25 Percentile	10.7
75 Percentile	43.4
P.E.	16.3
P.E. $_{\text{t.-obt. Av.}}$	.66

In Table X the gross gains are distributed in groups of 5. Beginning at the end of the table one sees that 2 children (or .3 per cent of the entire group) worked correctly from 96 to 100 combinations more in the final ten-minute period than in the initial ten-minute period; 2 children (or .3 per cent) worked from 91 to 95 more, etc. The distribution is somewhat skewed toward the high end. The range is from a loss of 19 combinations to a gain of 100 combinations. The median gain is 27.5 combinations. That is, the group as a whole, measured by the median gain, profited by the practice of fifty minutes to the extent that it worked correctly in 10 minutes at the end of practice 27.5 more combinations than it worked in 10 minutes at the beginning of the practice. This median gain means that there were as many children who gained 27.5 combinations or more as there were who gained 27.5 combinations or less. The upper 25 per cent of the class profited by the practice to the extent that they worked correctly in the final 10 minutes 43.4 combinations or more in excess of the number worked in the initial 10 minutes. The lower 25 per cent profited to the extent of 10.7 combinations or more. Only 25 of the 606 children failed to profit by the practice.

Relative Gain in Division

What relation did the gross gain from 50 minutes of practice bear to the initial ability is our next question for consideration. Or, putting it in other terms, what gain per cent in ability to make the division associations involved in this experiment resulted from 50 minutes of practice? This gain for the entire group was found from the individual gains made by the 606 children.

Table VII, column *17*, gives these gain per cents for one class. Boy A gained 30 per cent, boy B gained 78 per cent, boy C, 168 per cent, etc. The same points that were made in discussing the gain per cents in addition obtain here. These individual gain per cents were distributed class by class, boys and girls separate, but only the summary of this large table is given in Table XI, which shows the form of the distribution, the central tendency, and variability of the group.

In Table XI, the gain per cents in division combinations correctly answered resulting from 50 minutes of practice are distributed in groups of 15. The figures show that the distribution is skewed toward the high end. A wide range, from a loss of 74 per cent to a gain of 400 per cent, is seen. The mode is in the group 61 to 75 per cent. The median is 75 per cent. That is, there were just as many children who gained 75 per cent or more as there were who gained 75 per cent or less. The upper 25 per cent of the class gained 116 per cent or more, while the lower 25 per cent gained 47 per cent or less. The figures in the table show that 96 per cent of the children profited by the practice.

TABLE XI

GAIN PER CENT IN DIVISION COMBINATIONS CORRECTLY ANSWERED FROM FIFTY MINUTES OF PRACTICE

Gain, per cent....	-74 to -60	-59 to -45	-44 to -30	-29 to -15	-14 to 0	1 to 15	16 to 30	31 to 45	46 to 60	61 to 75	76 to 90
Individuals.......	1	3	2	5	14	16	42	59	77	85	76
Per cents........	.2	.5	.3	.8	2.3	2.6	6.9	9.7	12.7	14	12.5
Gain, per cent....	91 to 105	106 to 120	121 to 135	136 to 150	137 to 165	166 to 180	181 to 195	196 to 210	211 to 225	226 to 240	241 to 255
Individuals......	50	33	30	22	9	10	8	8	10	11	4
Per cents........	8.3	5.4	5	3.6	1.5	1.7	1.3	1.3	1.7	1.8	.7
Gain, per cent....	256 to 270	271 to 285	286 to 300	301 to 315	316 to 330	331 to 345	346 to 360	361 to 375	376 to 340	341 to 405	Total
Individuals.......	4	5	2	2	3	2	3	4	2	4	606
Per cents........	.7	.8	.3	.3	.5	.3	.5	.7	.3	.7	100

Median	75 per cent
25 Percentile	47
75 Percentile	116
P.E.	34.5

Gross Gain in Accuracy

To know the complete effect of the practice we must measure the change in accuracy of making the associations as well as the gain in speed in making them. The gain in accuracy is given as a gross amount, but is expressed in per cents. Table VII, column *18*, gives these gross gains for one class. They were found by subtracting the numbers in column *4* from the corresponding numbers in column *15*. Boy A lost 2 per cent, boy B gained 4 per cent, etc. The gains made by the 606 children were distributed class by class, girls and boys separate, but only a summary of the large table is given here, in Table XII, in which the form of the distribution, the central tendency and the variability of the group are clearly shown.

TABLE XII

GROSS GAIN IN ACCURACY IN DIVISION, EXPRESSED IN PER CENTS OF ANSWERS THAT WERE CORRECT. FROM FIFTY MINUTES OF PRACTICE

Per cent gained	–45 to –41	–40 to –36	–35 to –31	–30 to –26	–25 to –21	–20 to –16	–15 to –11	–10 to –6	–5 to –1	0 to 4
Individuals	1	1	0	2	3	3	11	37	113	211
Per cents	.2	.2	0	.3	.5	.5	1.8	6.1	18.6	34.8

TABLE XII—*Continued*

Per cent gained	5 to 9	10 to 14	15 to 19	20 to 24	25 to 29	30 to 34	35 to 39	40 to 44	45 to 49	50 to 54
Individuals	102	52	26	13	11	8	2	5	4	1
Per cents	16.8	8.6	4.3	2.1	1.8	1.2	.3	.8	.7	.2

Median	2.6	per cent
25 Percentile	1.4	
75 Percentile	8.	
P.E.	4.5	

Table XII shows a distribution of the gross gains in accuracy in division expressed in per cents of answers that were correct resulting from the 60 minutes of practice. The gains are given in groups of 5. The figures show a distribution conforming closely to the normal distribution curve. The range is from 45 per cent lost to 54 per cent gained. The median gain is 2.6 per cent which shows that there were as many children who gained 2.6 per cent or more in accuracy as there were who gained 2.6 per cent or less. The figures show that 435 children made the associations with equal or greater accuracy at the end of the practice than at the beginning, while 171 made them with less accuracy.

Combining now the evidence afforded by Tables X, XI, and XII, we can say that the 50 minutes of practice effected a very remarkable gain in the ability of these children to make the division associations concerned in the tests. Not only did the group as a whole increase its median ability to the extent that in 10 minutes at the end of practice it could do correctly 27.5 (or 75 per cent) more combinations than in 10 minutes at the beginning of the practice, but it also increased by 2.6 per cent its median accuracy in giving the answers.

Summary

ADDITION

1. 732 fourth-year children practiced for 75 minutes adding such columns as are shown on page 97. The gain from 60 minutes of the practice was measured.

2. Initial ability: In the initial 15 minutes of practice the median number of columns added correctly was 33.3.

3. Initial accuracy: In this initial 15-minute period, the median per cent of accuracy was 79 per cent.

4. Initial ability and accuracy: Expressing the initial ability by number of problems worked rather than by number of problems worked correctly, we can say that in the initial 15 minutes the group added a median number of 29.5 with a median accuracy of 79 per cent.

5. Absolute gain: In the final 15 minutes of practice the group added correctly a median of 10.7 more columns than in the initial 15-minute period. This gain resulted from 60 minutes of practice.

6. Relative gain: This gain of 10.7 columns meant a median percentile gain of 48 per cent, from 60 minutes of practice.

7. Change in accuracy: The change in accuracy was a median loss of .4 per cent, which was so small that it hardly needs to be considered in measuring the effect of the practice.

8. The central tendencies given are valid group measures only when considered with the wide deviations from them.

DIVISION

1. 606 children of last half of third year and first half of fourth year practiced for 60 minutes making such division combinations as are shown on page 98. The gain from 50 minutes of the practice was measured.

2. Initial ability: In the initial 10 minutes of practice the median number of combinations done correctly was 34.5.

3. Initial accuracy: In this initial 10-minute period, the median per cent of accuracy was 93 per cent.

4. Initial ability and accuracy: The initial ability may be expressed in number of combinations worked rather than in number worked correctly. The median number of combinations worked or attempted in 10 minutes was 37 with a median accuracy of 93 per cent.

5. Absolute gain: In the final 10 minutes of practice, the group worked correctly a median of 27.5 more combinations than in the initial 10 minutes of practice.

6. Relative gain: This gain of 27.5 combinations meant a median percentile gain of 75 per cent from 50 minutes of practice.

7. Change in accuracy: Along with the median gain of 75 per cent in speed went a median gain in accuracy of 2.6 per cent.

The Value of the Practice Experiment as a Method of Teaching

It would be easier to estimate the value of the practice experiment as a method for increasing skill if we had norms resulting from other methods of work with which to compare the results of this study. So far as the author knows, there has been no previous attempt to measure in a group of school children the progress resulting from the application of any particular method of school work. Mr. Courtis measured the change in a fifth-year grade from September to June in addition and division. However, the material used, the period of time for which the change was measured, and the method of work were all different, hence a comparison would confuse instead of clarify. Dr. Thorndike[6] has made a study of practice with university students in addition in which the same method and material used in this experiment were used. He found a median saving of time of 29 per cent, or improvement of 41 per cent in amount done per unit of time, the score used here. The time spent in practice amounted to about 53 minutes. He says: "The amount of improvement in this experiment may also add to our confidence that the method of the practice experiment wherein one works at one's limit and competes with one's own past record may well be made a regular feature in many school drills. Even if the same length of time produced in chil-

[6] Practice in the Case of Addition, *American Journal of Psychology,* Vol. XXI, pp. 483-488.

dren a percentile improvement only half as great as here, the gain would still probably be far greater than the gain by any of the customary forms of drill."

This is the nearest norm for a comparison known to the author. Dr. Thorndike's students practiced about 53 minutes; the children in the author's experiment practiced 60 minutes or about one and a fifth times as long. His students showed a median gain of 41 per cent; the children in this experiment, 48 per cent. In 50 minutes of practice in division the children in this experiment made a median gain of 75 per cent.

These gains, taken with Dr. Thorndike's statement that if children should make half as great a gain in the same time as was made by the university students in his experiment, they would gain more than by methods ordinarily employed in school, inspire confidence in the practice experiment as a most efficient method for conducting many of the drills necessary in regular school work.

J. C. Brown[7] in a study on the value of drill in arithmetic says, "Five-minute drill periods upon the fundamental number facts, preceding the daily lesson in arithmetic, were found to be beneficial in the sixth, seventh and eighth grades. . . . The benefit was not limited to improved mastery of the number habits, but included increased efficiency in arithmetical reasoning. The improvement was still in evidence after the lapse of the twelve weeks summer vacation." Mr. Brown does not describe the drill given except as follows: "The first five minutes of each period was devoted to drill work in addition, subtraction, multiplication and division. About one-half of the drill work was written and the other half oral." Mr. Brown provided for a control group. He found that the classes having this drill improved from 10 to 20 per cent more in different phases of arithmetic in the course of 20 recitations than the classes that did not have the practice. It is quite probable that even greater improvement than was found would have resulted had the drill been conducted by the use of practice experiment whereby each individual would have been competing with his own past record, thus having greater incentive to master thoroughly the facts involved.

---

[7] *Journal of Educational Psychology,* Feb. 1911, pp. 81 to 88, and Nov. 1912, pp. 485-492.

Factors Contributing to the Improvement

That a surprising improvement was made by these children working with the practice experiment as a method, has been conclusively proved by quantitative data. To try to determine some of the factors which contributed to this improvement is the present problem, a solution of which must rest upon conclusions more subjective in their nature. But the conclusions presented here are products of impressions made on the author while observing the reactions of the children to the work of this experiment; or of impressions from previous experience with children who studied by different methods the same facts involved in the experiment; or of impressions of the enthusiasm of children engaged in games where a perseverance in undergoing drudgery was displayed, that is seldom seen in the performance of school tasks, but whose presence in many phases of drill work would guarantee a most efficient and willing mastery of the facts involved. Any one who has watched children jump the rope, has observed their persistent efforts to outdo a rival, or to surpass their own previous performance. Boys will bounce a ball from the sidewalk against a wall and catch it with untiring zeal so long as there is a contest or a desire to surpass a previous record. It is quite noticeable in these games where the greatest persistence is displayed that children keep their record from time to time. It would be interesting to know if they would continue jumping the rope and tossing the balls with anything near the same enthusiasm if they knew nothing of their results from one trial to the next.

It was with the hope of finding a method for mastering some of the fundamental requirements of school work, that would at least make a partial appeal to this enthusiastic exuberance in children, that the practice experiment was tried, and the author feels that this appeal in the method used has been the greatest contributing factor to the gains made. However, there are many factors which may have entered into the gains made, some of which are here presented.

To what extent was the improvement due:

1. To increased power of concentrating on the task in hand?
2. To mastery of the technique of the experiment?

3. To greater incentive to effort in later tests than in the first one?
4. To work done outside of the time included in the test?
5. The lack of previous use of the functions involved?
6. To reinforcement of the mental activity by the motor activities involved in writing the results and by the visual percepts formed?
7. To maximum of opportunity afforded for exercising the function tested?
8. To the concerted effort of the group, each individual contributing a share to the gain?
9. To acquiring "higher order" habits of work?
10. To the freedom of the children to regulate their efforts for quantity and quality in the way best suited to their individual capacities for each?
11. To the stimulus resulting from knowing the exact score in quantity and quality of the previous test and to the desire to surpass this previous record?

The first five factors, if present to any great extent, would tend to produce an exaggerated measure of the gain in adding power and dividing power, or the gain found would be made up of the improvement in the functions tested plus the gain resulting from these five factors. The remaining six factors, it seems, would affect the measure of gain only as they directly influenced the ability to add and divide. An effort will be made to evaluate the influence of each of these eleven factors. To what extent was the improvement due to increased power of concentrating on the task in hand? In most experiments there is an increase in the ability to disregard conditions that make against the result sought. The children in this experiment acquired greater power of concentration and showed greater power to withstand distractions. How much this factor influenced can not be determined, but whatever the gain was it should be regarded as one of the valuable features of the method.

To what extent was the improvement due to mastering the technique of the experiment? In the addition experiment this factor certainly did not enter at all, since the children had been accustomed to add columns and place their answers beneath them. In the division the children had not been accustomed to writing short division in the form given on the sheets, but before the first recorded practice was made they were given enough practice in writing the results in this form to insure an understanding of its meaning. For a complete discussion of the care that was taken to insure this understanding on their part the reader is referred to Chapter I, page 6. There were a few children who could not get the results required by these sheets, but almost without exception these same children were unable to get the results any better by the customary form of division. Hence their failure or extreme slowness was due not to the form in which the division was expressed, but to their lack of knowledge of the process of division. Moreover, such records were not used in determining the gain. Further, the class records show no indication of misunderstanding this form of expressing division, since there is on the whole as great a gain between the performances of any two days as between those of the first and second. So the author feels certain that mastering the technique of this experiment produced no greater influence on the results, than this same factor ordinarily produces in an experiment.

To what extent was the improvement due to greater incentive to effort in later tests than in the first one? The children had greater incentive in the later tests because they knew their own previous records and were trying to surpass them. They had no such record to go against in the first test. Just how much this factor contributed to the results can not be told. Every other incentive was as great in the first test as in later ones.

To what extent was the improvement due to work outside of the time included in the test? As has been said before, some children worked at home to improve. This could not be prevented. Suggestions not to do so, might have resulted in more doing so. Very few did this; and these had to devise their own material, as no sheets were at their disposal. This factor, so far as it operated, tended to exaggerate the gains. How much can not be judged. While this uncontrolled factor detracts from the psychological value of the results, it indicates the value of the practice-experiment from the standpoint of education.

To what extent was the improvement due to lack of previous use of the function involved? The amount of improvement that may be expected of a group largely depends upon the place it has reached in the practice curve. Greater improvement may (at least in many cases) be expected in the early stages of the curve than in the later stages. One basis is available to show the comparative initial ability of the children in this group with that of other children in the same school grades. Three fourth-year classes containing 116 children in a school of another system using the same addition material worked a median of 21 columns with a median accuracy of about 75 per cent in 15 minutes. The children in the author's experiment worked a median of 30 columns with an accuracy of 79 per cent in the initial 15-minute period. While the initial ability for the three classes mentioned represents but a single school and so can not be considered as a norm of ability for fourth-year classes in general, it does indicate that the children in the present experiment had at least a fair degree of ability at the beginning of the practice and that there is no good reason for believing that the gains were unduly influenced by lack of previous use of the functions tested.

The five factors thus far discussed would tend in general to exaggerate the improvement made to the degree that they operated in the group tested. However, the author feels that their influence was very slight. The following six factors are the ones which best account for the improvement made.

To what extent was the improvement due to the reinforcement of the mental activity by the motor activity involved in writing the results and by the visual percepts formed of the results? This opportunity for improvement was present at least in the division to a much greater extent than is true in most practice given in school. Formulating the results so that they could be written required a complete decision which might be shirked in a method not demanding a written answer. Writing the results required motor responses which in turn left their impression upon the mind. The results when written were objective forms to make further lasting impression through the sense of sight. The author attributes much of the gain in division to the writing of the results which demanded sharper thinking, induced motor activity, and made possible more acute visual images.

To what extent was the improvement due to the maximum of opportunity afforded by the practice for exercising the function tested? Any one who has observed classes drilling by the methods commonly used in school knows that much of the time is consumed in the manipulation of cards, in writing numbers preparatory for the drill, or in mere dawdling. In the method of drill used in the experiments of this study, almost all of the energy expended was continually directed toward making the associations whose perfecting the drill expected to accomplish. There was a minimum of writing to be done, which demanded a maximum of thinking, and which also tended to keep the mind on the work to be done. One is quite certain that in ten minutes of drill with this method, children had the opportunity to think and express many more results than with methods ordinarily used in school. Greater opportunity for concentrating on the facts to be fixed in mind could hardly be devised. So we may be certain that much of the gain was due to the children's making very many associations within a short time, which provided for frequent repetition of associations, one of the fundamental requisites for perfect habit formation.

To what extent was the improvement due to the concerted effort of the group, each individual contributing a share to the gain? A prominent element in determining the amount of improvement in any group is the part of the group that contributes to the improvement. Effort distributed throughout the group gives expectation of greatest improvement. The author has never seen children work when there was as great an effort on as great a part of a group.

To what extent was the improvement due to acquiring "higher order" habits of work? Very many children in the third and fourth years of school form their combinations, especially in addition, by counting. No effort was made to get children to change their methods. It is not easy to tell with certainty just what children use the lower habits and which ones add by use of the combinations. No effort was made to keep a record of individuals' methods. This could hardly be done because even children who are most addicted to adding by ones, add on the smaller numbers, like two and three, at one step. So, to make an exact classification of the children on this basis would be impossible even if one could tell just what method a child were using at any instant. However, if children do acquire the higher order habits of adding by gradually taking up new combinations, it is reasonable to assume that, with the desire these children had to advance, many of them acquired higher order habits of work which contributed a share to the gain that was made. On the other hand, this changing to a new method, in some instances that were very noticeable, caused an increase of errors that resulted in a loss. It is believed by the author that many of the individual gross losses were due to an unusual increase of errors resulting from some such change in method of work. Still for the group as a whole it is reasonable to believe that some share of the gain was due to the acquisition of higher order habits of work.

To what extent was the improvement due to the freedom of the children to regulate their efforts for quantity and quality in the way best suited to their individual capacities for each? In most school work such stress is placed on quality of work that a child not only is not rewarded for quantity of work, but is often made uncomfortable for excessive production regardless of whether he has maintained his standard of quality. Individuals vary greatly in speed and accuracy of work. Rapid calculators are as a rule more accurate than slow ones. However, this does not mean that by speeding the slow one up to the fast one's norm he would become more accurate; but there is considerable evidence to show that decreasing the speed of the fast one to the norm of the slow one would improve his accuracy little or not at all. In these tests children were left free within certain limits to regulate their effort on quantity and quality in such a way as to produce the highest score when the penalties for inaccuracy had been assessed. Most of the children soon saw that they could work more rapidly without increasing their errors and regulated their efforts accordingly. The fact that the group decreased only .4 of one per cent in median accuracy in addition and increased 2.6 per cent in median accuracy in division while there was a median gain per cent in speed of 50 per cent, shows conclusively that quantity of work may be increased to a vast extent without deterioration in its quality. To just what extent the improvement was due to the way the children regulated their effort for quantity and quality can not be told, but one may say with assurance that the improvement would have been much less had the children followed the usual maxim, "Not how much but how well."

To what extent was the improvement due to the stimulus resulting from knowing the exact score in quantity and quality of the previous test and to a desire to surpass this previous record? The author feels that the children's knowledge of the record of their previous performance, united with the desire each one had to surpass his previous record, was the greatest contributing factor to the improvement. Concerning the value of this factor Ladd and Woodworth[8] say: "Experimental conditions are stimulating largely because one has a measure of one's success and progress; and the habit of checking up one's work can scarcely fail to prove of benefit wherever measures of success and failure are practicable." Not only were the children anxious to know the results of their previous tests, but they exerted such a determined effort each to surpass his own previous record, that one felt progress must necessarily result. At the end of a test there were frequent expressions of regret that the time was up, but almost no indication of indifference or ennui. The children were deriving immediate satisfaction from winning in a game where each was striving to be a victor, not under the enervating condition that he try to surpass someone who is natively his superior, but under the energizing condition that he surpass his own previous day's performance. This factor provided incentive for the slow child who ordinarily realized the futility of striving to do as well as the exceptional members of the class; it provided an incentive also for these exceptional members to exert their best effort, and thus precluded the dawdling that many methods cultivate among the more gifted members of a class. So, under the conditions of this game of practice, most of the children were successful, and "nothing succeeds like success." Ladd and Woodworth[9] appraise the emotional tone attending successful and baffled effort as follows: "Exactly how these emotions act to strengthen one association and to weaken or counteract another cannot readily be seen; but it is safe to assume that they corresponded to some genuine dynamic process of great efficacy." So, much of the gain in efficiency must have been due to the children's knowing their previous success, which in turn induced a pleasurable emotional tone to inspire continued intense effort and strengthen associations formed.

The intense effort that was displayed, the improvement that was made, and the emotional attitude that was aroused, seem to

[8] Elements of Physiological Psychology, p. 571.

[9] *Ibid.*, p. 552.

be features that should recommend the practice experiment as a method for school work. Bryan and Harter[10] say: "A school method must be judged by the moods and tempers which it cultivates not simply by what is learned, still less by the momentary interest it arouses. If one forces mastery of the multiplication tables by a method which keeps one half the school cowed and the other half rebellious, one has obtained a useful result but at disastrous cost. Better not know the multiplication tables than be thus morally maimed. There are many schools and homes where hard tasks are performed in a good temper; where thorough drill does not arrest, but prepares the way for higher development, where children begin to do what they must later do to succeed in any business—pass cheerfully from interest in desired ends to a resolute drudgery necessary for the attainment of these ends." That the practice experiment, wherever applicable, meets the demands made here of a method for school work, is confidently asserted.

[10] *Psych. Rev.*, Vol. I, p. 370.

CHAPTER III

THE EFFECT OF THE DISTRIBUTION AND LENGTH OF WORK PERIOD UPON THE RATE OF LEARNING

The problem of this chapter is to measure the effect of giving about one hour of drill in arithmetic in periods of different length. If we have an hour to devote to drill in the division tables, or multiplication tables or in addition, is it better to give this drill in short periods extending over a correspondingly greater number of days, or is it better to give the drill in longer periods thus concentrating it into fewer days?

Plan of the Practice

To understand the manner in which the solution to this problem was attempted, it is necessary to recall the plan of giving the tests that have been discussed in the preceding chapters. In the addition experiment with fourth-year children, there were four groups of classes on the basis of the distribution of the practice. In all the groups there was an initial practice-period and a final practice-period of fifteen minutes each, but the intervening forty-five minutes were distributed differently for the different groups. In the first group the intervening forty-five minutes were divided into two practice-periods of twenty-two and a half minutes each; in the second group into three practice-periods of fifteen minutes each; in the third group into seven practice-periods of six minutes each and one period of three minutes; in the fourth group into twenty-one two-minute periods and one three-minute period. The following plan makes this distribution clear:

Groups	Initial Period	Intervening 45 Minutes		Final Period
I	15 min.	2	22½ min.	15 min.
II	15 min.	3	15 min.	15 min.
III	15 min.	7	6 min. and 1 3 min.	15 min.
IV	15 min.	21	2 min. and 1 3 min.	15 min.

Each group of classes practiced once a day on successive school days, as far as possible,[11] which of course resulted in the experiment extending over a different number of days for the different groups. For Group I, four successive days were required; for Group II, five successive days; for Group III, ten successive days; and for Group IV, twenty-four successive days.

TABLE XIII

NUMBER OF PROBLEMS ADDED CORRECTLY IN THE INITIAL FIFTEEN-MINUTE PERIOD BY THE FOUR GROUPS

(EACH DESIGNATED BY THE NUMBER OF MINUTES IN ITS INTERVENING PRACTICE-PERIODS)

Columns added correctly	Group I		Group II		Group III		Group IV	
	22½ min.		15 min.		6 min.		2 min.	
	Individuals	Per cents	Individuals	Per cents	Individuals	Per cents	Individuals	Per cents
0 to 4					1	.5	2	.9
5 to 9	16	8.2	6	5.8	10	4.9	18	7.9
10 to 14	23	11.9	9	8.7	35	17.1	13	5.7
15 to 19	32	16.5	16	15.4	45	22	40	17.4
20 to 24	38	19.6	19	18.3	36	17.6	37	16.2
25 to 29	27	13.9	11	10.6	33	16.1	40	17.4
30 to 34	19	9.8	14	13.5	19	9.2	29	12.7
35 to 39	13	6.7	12	11.5	11	5.4	21	9.2
40 to 44	7	3.6	4	3.8	8	3.9	14	6.1
45 to 49	2	1.	4	3.8	4	2.	4	1.8
50 to 54	5	2.6	2	1.9	2	1.	4	1.8
55 to 59	4	2.1	0	0	0	0	3	1.3
60 to 64	2	1.	1	1.	0	0	1	.5
65 to 69	1	.5	0	0	1	.5	1	.5
70 to 74	4	2.1	2	1.9			1	.5
75 to 79	0	0	0	0			0	0
80 to 84	0	0	3	2.9			0	0
85 to 89	1	.5					0	0
90 to 94							1	.5
95 to 99								
100 to 104								
105 to 109			1	1				
Total	194	100	104	100	205	100	229	100
Average	25.9		29.2		22.7		26.3	
Median	22.9		25.4		21.1		25.1	
25 P.	16.		17.9		15.1		17.5	
75 P.	32.		35.8		28.6		33.3	
P.E.	8.		8.9		6.8		7.9	
$P.E._{t.\text{-}obt.\ Av.}$	.6		.9		.5		.5	

[11] For exceptions and further details see Chapter I, "Plan of the Practice."

In Group I, there were 6 classes, containing 194 children; in Group II,[12] 3 classes, containing 104 children; in Group III, 6 classes, containing 205 children; and in Group IV, 6 classes, containing 229 children.

Addition

Initial Ability of the Groups

The classes to compose any group were chosen at random, or sometimes on the basis of convenience in reaching them. Since nothing was known of the exact ability of the classes in addition before beginning the tests, this plan of grouping seemed justifiable. Had it been possible to have the tests for all these classes in progress at one time, one could have grouped them on basis of ability, after the first practice-period, in such a way as to secure equality of initial ability in the four groups. Table XIII shows a summary of the distribution of the initial scores of all these classes grouped, as has been shown above, on the basis of the length of the practice-periods for the 45 minutes of practice between the initial practice-period and the final practice-period. The data composing this table are the same as those composing Table II, but they are distributed here to show the initial ability of the four groups that are being compared. The table shows the number of problems worked correctly in the initial 15-minute practice-period by the members of each group. Samples of the scores distributed here may be seen in Table I, column *3*.

Table XIII reads as follows: In Group I, 16 children (or 8.2 per cent of the group) added from 5 to 9 columns correctly in the initial 15-minute period; in Group II, 6 children (or 5.8 per cent of the group) showed the same ability; in Group III, 10 children (or 4.9 per cent of the group) showed the same ability; and in Group IV, 18 children (or 7.9 per cent of the group) showed the same ability. The range of the four groups was slightly different, but not enough to deserve much weight. The central tendencies as shown by the averages of the groups

[12] After the experiment had progressed for some time it seemed more profitable to discontinue giving the 15-minute practice-periods, since they differed so little from the 22½-minute periods, and to concentrate effort on the extreme and middle periods. This accounts for the smaller number of classes and children in Group II, which in turn causes less weight to be given to the measures of Group II in the conclusions.

were 25.9, 29.2, 22.7 and 26.3. By medians they were 22.9, 25.4, 21.1 and 25.1. The individuals comprising the middle 50 per cent of the groups had the following ranges: Group I, 16 columns to 32 columns; Group II, 18 columns to 36 columns; Group III, 15 columns to 29 columns; Group IV, 18 columns to 32 columns. According to the percentages given in the table, about 44 per cent of Group I, 52 per cent of Group II, 38 per cent of Group III, and 52 per cent of Group IV, added correctly 25 columns or more.

TABLE XIV

PER CENT OF PROBLEMS ADDED CORRECTLY IN THE INITIAL FIFTEEN-MINUTE PERIODS BY THE FOUR GROUPS

(EACH DESIGNATED BY THE NUMBER OF MINUTES IN ITS INTERVENING PRACTICE-PERIOD)

Per cent of problems correct	Group I		Group II		Group III		Group IV	
	22½ min.		15 min.		6 min.		2 min.	
	Individuals	Per cents	Individuals	Per cents	Individuals	Per cents	Individuals	Per cents
6 to 10								
11 to 15			1	1			1	.5
16 to 20							0	
21 to 25	1	.5					1	.5
26 to 30	2	1.					1	.5
31 to 35	2	1.	1	1.	2	1.	4	1.8
36 to 40	6	3.1	1	1.	2	1.	4	1.8
41 to 45	5	2.6	2	1.9	5	2.5	4	1.8
46 to 50	4	2.1	2	1.9	7	3.4	9	3.9
51 to 55	6	3.1	0	0	4	2.	7	3.1
56 to 60	7	3.6	1	1.	10	4.9	7	3.1
61 to 65	21	10.9	5	4.8	9	4.4	11	4.8
66 to 70	14	7.2	7	6.7	16	7.8	21	9.2
71 to 75	13	6.7	17	16.4	28	13.2	26	11.3
76 to 80	22	11.4	12	11.5	28	13.2	38	16.7
81 to 85	31	16.	11	10.6	16	7.8	23	10.1
86 to 90	27	13.9	17	16.4	27	13.2	37	16.2
91 to 95	19	9.8	14	13.5	21	10.3	22	9.6
96 to 100	14	7.2	13	12.5	30	14.6	13	5.7
Total	194	100	104	100	205	100	229	100
Median	79		82		79		78	
25 P.	64		72		69		68	
75 P.	88		91		90		88	
P.E.	12		9.5		10.5		10	

In order to give a more exact statement of the relative initial ability of these four groups in addition, it is necessary to show the accuracy with which each group did its work. In finding the accuracy of the groups, the same data are used as were used in Table III, but now the summaries show the distributions for the classes grouped according to the length of the practice-periods which comprised the intervening 45 minutes of practice. This distribution is found in Table XIV. Samples of the individual records which comprise this table may be found in Table I, column *4*.

The medians for accuracy are: for Group I, 79 per cent; for Group II, 82 per cent; for Group III, 79 per cent; and for Group IV, 78 per cent. This means that 50 per cent of Group I added with an accuracy of 79 per cent or more; 50 per cent of Group II added with an accuracy of 82 per cent or more; 50 per cent of Group III added with an accuracy of 79 per cent or more, etc.

In terms of number of problems worked and accuracy the following is the initial status of each group; Group I added 29 columns with an accuracy of 79 per cent. Group II added 31 columns with a median accuracy of 82 per cent. Group III added 27 columns with a median accuracy of 79 per cent. Group IV added 32 columns with an accuracy of 78 per cent. The median initial ability of the groups summarized is as follows:

	Median Number of Columns Marked	Median Number of Columns Correct	Accuracy
Group I	29 columns	22.9	79 per cent
Group II	31 columns	25.4	82 per cent
Group III	27 columns	21.1	79 per cent
Group IV	32 columns	25.1	78 per cent

Gross Gain

Having defined the initial ability of the four groups in terms of quality and quantity of performance, we shall now determine the absolute gain made by each group in the course of 75 minutes of practice, 30 minutes of which were occupied by the initial 15-minute practice-period and the final 15-minute practice-period for all groups alike; the remaining 45 minutes of which were divided into practice-periods of 22½ minutes for Group I, 15 minutes for Group II, 6 minutes for Group III, and 2 minutes for Group IV.

The data to be used in reaching a conclusion to this problem are the same as those used in Table IV, the source of which was described minutely in connection with that table. In Table

XV the gross gains made by the 732 children, samples of which may be seen in Table I, column *14*, are distributed for the four groups.

TABLE XV

GROSS GAIN IN NUMBER OF COLUMNS ADDED CORRECTLY IN FIFTEEN MINUTES, MADE IN THE COURSE OF SEVENTY-FIVE MINUTES OF PRACTICE BY THE FOUR GROUPS

Columns gained	Group I		Group II		Group III		Group IV	
	22½ min.		15 min.		6 min.		2 min.	
	Individuals	Per cents	Individuals	Per cents	Individuals	Per cents	Individuals	Per cents
–15 to –12	2	1.	2	1.9			1	.5
–11 to –8	0	0	4	3.8	5	2.5	2	.9
–7 to –4	8	4.1	4	3.8	11	5.4	5	2.3
–3 to 0	12	6.2	8	7.7	29	14.1	14	6.2
1 to 4	35	18.	9	8.7	21	10.3	28	12.2
5 to 8	34	17.5	10	9.6	28	13.2	33	14.4
9 to 12	25	12.9	24	23.1	31	15.1	31	13.6
13 to 16	27	13.9	12	11.5	22	10.8	29	12.7
17 to 20	22	11.3	7	6.7	25	12.2	21	9.2
21 to 24	4	2.1	4	3.8	9	4.4	12	5.4
25 to 28	10	5.2	3	2.9	7	3.4	9	2.9
29 to 32	6	3.1	3	2.9	9	4.4	7	3.1
33 to 36	6	3.1	3	2.9	2	1.	13	5.7
37 to 40	1	.5	5	4.8	3	1.5	7	3.1
41 to 44	0	0	1	1.	1	.5	3	1.3
45 to 48	1	.5	2	1.9	1	.5	3	1.3
49 to 52	1	.5	1	1.	0	0	3	1.3
53 to 56			0	0	0	0	2	.9
57 to 60			0	0	1	.5	2	.9
61 to 64			2	1.9			2	.9
65 to 68							2	.9
Total	194	100	104	100	205	100	229	100
Average	11.		13.6		10.7		16.1	
Median	9.5		11.		9.6		12.6	
25 P.	3.5		4.1		1.7		5.4	
75 P.	17.		19.4		17.6		23.1	
P.E.	6.8		7.7		8.		8.9	
P.E.$_{\text{t.-obt. Av.}}$	.5		.8		.6		.6	

Table XV reads as follows in the fifth line: "In Group I, 35 children (or 18 per cent of the group) gained from 1 to 4 problems; in Group II, 9 children (or 8.7 per cent of the group) gained from 1 to 4 problems; in Group III, 21 children (or 10.3 per cent of the group) gained from 1 to 4 problems; in Group IV, 28 children (or 12.2 per cent of the group) gained from 1 to 4 problems." The range of distribution is somewhat wider for the two-minute group than for the other three groups, being from 15 columns lost to 68 gained. The four distributions have the same general form, being slightly skewed at the high end. The median gain for the 22½-minute group is 9.5 columns; for the 15-minute group, 11 columns; for the 6-minute group, 9.6 columns; and for the 2-minute group, 12.6 columns. The respective averages were 11.0, 13.6, 10.7 and 16.1. These medians and averages show an advantage for the group which had the forty-five minutes of practice in 2 minute periods. The 22½-minute group and the 6-minute groups made about the same gain, while the 15-minute group made a larger gain. The percentages in the table show that:

40 per cent of Group I gained 13 columns or more;
41 per cent of Group II gained 13 columns or more;
39 per cent of Group III gained 13 columns or more;
50 per cent of Group IV gained 13 columns or more.

Percentile Gain

To get a measure of the gain of each group in terms of its initial performance is our next problem. To find the gain per cent of each group of classes, the gain per cents made, by all the individuals, samples of which may be seen in Table I, column *15*, have been distributed in four groups, formed on the basis previously described. These data are given in Table XVI.

Table XVI reads as follows in the seventh line: " In Group I, 23 children (or 11.9 per cent of the group) gained from 1 to 15 per cent in columns added correctly; in Group II, 10 children (or 9.6 per cent of the group) gained the same; in Group III, 15 children (or 7.5 per cent of the group) gained the same; in Group IV, 18 children (or 7.9 per cent of the group) gained the same." The range of the groups is from 89 per cent lost to more than 300 per cent gained, that of each group being about the same. The form of the distribution for each group is similar, being skewed toward the high end.

The median gain per cents for the groups made in the course of 75 minutes of practice are as follows: Group I, 45 per cent, Group II, 43 per cent; Group III, 42 per cent, and Group IV, 56 per cent.

TABLE XVI

GAIN PER CENT IN COLUMNS ADDED CORRECTLY MADE IN THE COURSE OF SEVENTY-FIVE MINUTES PRACTICE BY THE FOUR GROUPS

Gain, per cent	Group I		Group II		Group III		Group IV	
	22½ min.		15 min.		6 min.		2 min.	
	Individuals	Per cents	Individuals	Per cents	Individuals	Per cents	Individuals	Per cents
–89 to –75	1	.5	2	1.9				
–74 to –60	0	0	1	1.				
–59 to –45	0	0	1	1.	3	1.5	2	.9
–44 to –30	2	1.	2	1.9	7	3.4	0	0
–29 to –15	4	2.1	4	3.8	5	2.5	7	3.1
–14 to 0	11	5.7	8	7.7	29	13.7	13	5.7
1 to 15	23	11.9	10	9.6	15	7.5	18	7.9
16 to 30	29	14.9	12	11.6	27	13.2	29	12.7
31 to 45	28	14.2	14	13.4	21	10.3	22	9.6
46 to 60	29	14.9	9	8.7	20	9.8	34	14.9
61 to 75	24	12.4	11	10.6	15	7.5	30	12.1
76 to 90	15	7.8	6	5.8	16	8.	17	7.4
91 to 105	8	4.1	5	4.8	9	4.4	18	7.9
106 to 120	5	2.6	5	4.8	10	4.9	9	3.9
121 to 135	5	2.6	3	2.9	3	1.5	3	1.3
136 to 150	3	1.5	5	4.8	8	3.9	5	2.3
151 to 165	2	1.	1	1.	2	1.	4	1.8
166 to 180	1	.5	0	0	1	.5	5	2.3
181 to 195	1	.5	0	0	1	.5	0	0
196 to 210	0	0	2	1.9	4	2.	1	.5
211 to 225	1	.5	0	0	1	.5	2	.9
226 to 240	1	.5	1	1.	3	1.5	3	1.3
241 to 255	0	0	1	1.	0	0	0	0
256 to 270	0	0			0	0	0	0
271 to 285	0	0			1	.5	1	.5
286 to 300	0	0			2	1.	0	0
301 to 305	1	.5	1	1.	2	1.	6	2.7
Total	194	100	104	100	205	100	229	100
Median	45		43		42		56	
25 P.	19		13		8		24	
75 P.	72		86		87		90	
P.E.	26.5		26.5		39.5		33	

Gain in Accuracy

One other factor remains to be considered in measuring the effect of the practice on the four groups, the change in accuracy wrought by the practice. The data to determine this result have been used before in Table VI. The source of these data may be seen by referring to Table I, column *16*. A distribution of the gains in accuracy for the four groups is given in Table XVII.

TABLE XVII

GROSS GAIN IN ACCURACY IN ADDITION, EXPRESSED IN PER CENTS OF ANSWERS THAT WERE CORRECT, MADE IN THE COURSE OF SEVENTY-FIVE MINUTES OF PRACTICE BY THE FOUR GROUPS

Per cents gained	Group I 22½ min.		Group II 15 min.		Group III 6 min.		Group IV 2 min.	
	Individuals	Per cents	Individuals	Per cents	Individuals	Per cents	Individuals	Per cents
–55 to –51			1	1.	2	1.	5	2.3
–50 to –46			2	1.9	0	0		
–45 to –41			0	0	2	1.		
–40 to –36			2	1.9	4	2.		
–35 to –31	2	1.			12	5.9	5	2.3
–30 to –26	2	1.	5	4.8	5	2.5	11	4.8
–25 to –21	3	1.5	3	2.9	4	2.	7	3.1
–20 to –16	9	4.5	5	4.8	20	9.8	20	8.7
–15 to –11	12	6.2	7	6.7	21	10.3	23	10.1
–10 to –6	17	8.7	12	11.6	19	9.2	20	8.7
–5 to –1	23	11.9	19	18.3	17	8.2	42	18.4
0 to 4	36	18.5	19	18.3	25	12.2	31	13.6
5 to 9	26	13.4	8	7.7	22	10.8	23	10.1
10 to 14	23	11.9	7	6.7	16	7.8	15	6.6
15 to 19	15	7.8	6	5.8	15	7.3	11	4.8
20 to 24	10	5.2	3	2.9	10	4.9	5	2.3
25 to 29	4	2.1	3	2.9	5	2.5	5	2.3
30 to 34	3	1.5	0	0	2	1.	1	.5
35 to 39	5	2.6	0	0	1	.5	2	.9
40 to 44	1	.5	1	1	0	0		
45 to 49	2	1.	1	1.	2	1.		
50 to 54					0	0	1	.5
55 to 59					1	.5	2	.9
60 to 64								
65 to 69	1	.5						
Total	194	100	104	100	205	100	229	100
Median	3.5		–1.6		–1.5		–2.7	
25 P.	–4.7		–10.1		–15		–13.5	
75 P.	12.9		6.4		9.7		6.2	
P.E.	8.8		8.3		12.4		9.8	

Table XVII reads as follows near the middle: "In Group I, 36 children (or 18.5 per cent of the group) gained from 0 to 4 per cent in accuracy; in Group II, 19 children (or 18.3 per cent of the group) gained the same; in Group III, 25 children (or 12.2 per cent of the group) gained the same; in Group IV, 31 children (or 13.6 per cent of the group) gained the same."

The median change in accuracy in the four groups is as follows: Group I gained 3.5 per cent; Group II lost 1.6 per cent; Group III lost 1.5 per cent; Group IV lost 2.7 per cent.

Summary

The following summarized statement gives the data necessary for comparing the four groups and for interpreting the results.

MEDIANS OF THE GROUPS OF INDIVIDUALS

	Median Initial Ability: Examples Correct	Average Gross Gain	Median Gross Gain	Reliability: P.E.$_{\text{t.-obt. Av.}}$	Median Gain Per cent	Median Gain in Accuracy
Group I	22.9	11.0	9.5	.5	45	3.5
Group II	25.4	13.6	11.0	.8	43	-1.6
Group III	21.1	10.7	9.6	.6	42	-1.5
Group IV	25.1	16.1	12.6	.6	56	-2.7

In computing all averages and medians thus far, the individual in the group has been considered as the unit. We shall now present the same data computed, with the *class* as the unit. The *median* ability of each class was computed; then the *average* of these *class-medians* for each group was found. The reliability of this average was computed on the basis of the number of classes. The following figures show the result.

AVERAGE OF CLASS MEDIANS

	Average Initial Median Ability of Classes: Examples Correct	Reliability of Average: P.E.$_{\text{t.-obt. Av.}}$	Average Gross Median Gain of Classes	Reliability of Average: P.E.$_{\text{t.-obt. Av.}}$	Average of Median Percentile Gains
Group I	23.7	1.7	10.2	1.1	42
Group II	25.7	2.3	9.6	1.2	41
Group III	21.3	.9	9.4	.8	42
Group IV	25.5	1.5	13.9	1.9	58

Using now the best measure of efficiency—the number of examples correct, which includes credit for both speed and accuracy—and the three methods of computing the gain, we have:

	AVERAGE GROSS GAIN OF INDIVIDUALS	MEDIAN GROSS GAIN OF INDIVIDUALS	AVERAGE OF MEDIAN GROSS GAINS OF CLASSES
Group I	11.0	9.5	10.2
Group II	13.6	11.0	9.6
Group III	10.7	9.6	9.4
Group IV	16.1	12.6	13.9

It appears that, taking the results at their face value, the 2-minute practice was the most favorable. Part of its superiority to the 22½-minute and 6-minute practice was probably due to the greater initial ability of those taking it. This will be studied later. Part of its superiority may have been due also to chance factors, but these are as likely to have worked against as for it.

Part of its superiority may have been due to the greater length of time covered, and so the greater amount of opportunity for training in and out of school, apart from the special practice. We shall return to this after surveying the facts concerning division.

Division

Initial Ability

In the division experiment, with children in the last half of the third year and the first half of fourth year, there were three groups of classes on the basis of the distribution of the 60 minutes of practice. In all of the groups there was an initial practice-period and a final practice-period of 10 minutes each, but the intervening 40 minutes were distributed differently for the different groups. In Group I, the intervening 40 minutes of practice were divided into 2 practice-periods of 20 minutes each; in Group II, into 4 practice-periods of 10 minutes each; and in Group III, into 20 practice-periods of 2 minutes each. The following plan shows clearly this distribution of the practice.

Groups	Initial Practice-Period	Intervening 45 Minutes of Practice	Final Practice-Period
I	10 min.	2 20 min. periods	10 min.
II	10 min.	4 10 min. periods	10 min.
III	10 min.	20 2 min. periods	10 min.

The classes of each of the groups practiced once a day on successive school days as far as possible,[13] which of course resulted in the experiment extending over a different number of days for the different groups. For Group I, 4 successive days; for Group II, 6 successive days; and for Group III, 22 successive days were required to complete the experiment. In Group I, there were 6 classes containing 204 children; in Group II, 6 classes containing 209 children; and in Group III, 6 classes containing 193 children. The classes to compose the different groups were chosen at random or on the basis of accessibility for the author, since nothing definite was known of their ability in division before beginning the tests to serve as a basis for grouping whereby equality of initial ability might be secured in the different groups.

Table XVIII shows a distribution of the number of combina-

[13] For exceptions and further details see Chapter I, "Plan of the Practice."

tions worked correctly in the initial 10-minute period of practice by the 606 children, arranged in three groups made on the basis of the length of the intervening periods of practice. Samples of these scores may be seen in Table VII, column *3*.

TABLE XVIII

The Number of Division Combinations Answered Correctly in the Initial Ten-minute Period by Each of the Three Groups

Number of combinations	Group I		Group II		Group III	
	20 min.		10 min.		2 min.	
	Individuals	Per cents	Individuals	Per cents	Individuals	Per cents
5 to 9	10	4.9	10	4.8	3	1.6
10 to 14	15	7.4	12	5.7	6	3.1
15 to 19	19	9.3	22	10.5	24	12.4
20 to 24	13	6.4	24	11.5	19	9.8
25 to 29	15	7.4	26	12.4	11	5.7
30 to 34	15	7.4	21	10.	15	7.8
35 to 39	20	9.8	21	10.	16	8.3
40 to 44	17	8.3	12	5.7	16	8.3
45 to 49	12	5.9	19	9.1	12	6.7
50 to 54	18	8.8	12	5.7	8	4.1
55 to 59	14	6.9	10	4.8	18	9.3
60 to 64	16	7.8	5	2.4	7	3.6
65 to 69	4	2.	8	3.8	3	1.6
70 to 74	2	1.	3	1.4	8	4.7
75 to 74	8	3.9	3	1.4	5	2.6
80 to 84	0	0	0		3	1.6
85 to 89	2	1.	1	.5	6	3.1
90 to 94	1	.5			4	2.1
95 to 99	3	1.5			2	1.
100 to 104					0	0
105 to 109					3	1.6
110 to 114					1	.5
115 to 119					0	0
120 to 124					1	.5
125 to 129					1	.5
Total	204	100	209	100	193	100
Average	39.1		34.6		44.6	
Median	38.3		32.		40.3	
25 P.	22.2		21.2		23.5	
75 P.	54.2		46.8		58.3	
P.E.	16		12.8		17.4	
P.E.$_{\text{t.-obt. Av.}}$	1.1		.9		1.3	

The average number of combinations worked correctly was: for Group I, 39.1; for Group II, 34.6; for Group III, 44.6. The median number of combinations worked correctly was: for

Group I, 38.3; for Group II, 32; for Group III, 40.3. The per cents in the table show that 47 per cent of Group I, and 35 per cent of Group II, reached the median of Group III. In evaluating the gains made by the three groups, this difference in initial ability must be taken into consideration.

The accuracy of the three groups in doing the division combinations is another factor which must be determined in order to show exactly the comparative ability of the groups in their initial performance. The data to be used are the per cents of accuracy for the 606 children, samples of which may be seen in Table III, column *4*. These data are distributed in Table XIX for the three groups.

TABLE XIX

PER CENT OF CORRECT ANSWERS TO THE DIVISION COMBINATIONS IN THE INITIAL TEN MINUTES BY THE THREE GROUPS

Per cent of correct answers	Group I		Group II		Group III	
	20 min.		10 min.		2 min.	
	Individuals	Per cents	Individuals	Per cents	Individuals	Per cents
26 to 30			1	.5		
31 to 35						
36 to 40			1	.5		
41 to 45	2	1.	1	.5		
46 to 50	2	1.	3	1.4	2	1.
51 to 55	3	1.5	2	1.		
56 to 60	2	1.	4	2.	1	.5
61 to 65	6	3.	3	1.4	3	1.6
66 to 70	5	2.5	3	1.4	3	1.6
71 to 75	6	3.	11	5.3	5	2.6
76 to 80	6	3.	15	7.2	10	5.2
81 to 85	11	5.4	15	7.2	27	14.
86 to 90	22	10.8	27	12.9	21	10.9
91 to 95	47	23.	57	27.2	42	21.7
96 to 100	92	45.1	66	31.6	79	40.9
Total	204	100	209	100	193	100
Median	94		92		93	
25 P.	87		83		85	
75 P.	98		97		98	
P.E.	5.5		7		6.5	

Table XIX reads as follows at the bottom: In Group I, 92 children (or 45.1 per cent of the group) worked the combinations in the initial practice-period with an accuracy of 96 to 100 per cent; in Group II, 66 children (or 31.6 per cent of the group) worked with the same accuracy; and in Group III, 79 children (or 40.9 per cent of the group) worked with the same accuracy. The three distributions are very similar, being decidedly skewed toward the low end. The median accuracy for the three groups is almost equal. Group I worked with a median initial accuracy of 94 per cent, Group II 92 per cent, and Group III 93 per cent. So in the factor of accuracy there is not enough difference to demand evaluation.

TABLE XX

NUMBER OF DIVISION COMBINATIONS CORRECTLY ANSWERED GAINED IN THE COURSE OF SIXTY MINUTES OF PRACTICE BY THE THREE GROUPS

Number of combinations	Group I		Group II		Group III	
	20 min.		10 min.		2 min.	
	Individuals	Per cents	Individuals	Per cents	Individuals	Per cents
-19 to -15	1	.5	1	.5		
-14 to -10	2	1.	3	1.4		
-9 to -5	0	0	2	1.	1	.5
-4 to 0	8	3.9	6	2.8	1	.5
1 to 5	10	4.9	5	2.5	5	2.6
6 to 10	17	8.3	25	11.9	7	3.6
11 to 15	27	13.2	18	8.6	10	5.2
16 to 20	25	12.3	29	13.9	6	3.1
21 to 25	29	14.2	26	12.4	14	7.3
26 to 30	23	11.3	20	9.5	17	8.8
31 to 35	18	8.8	19	9.1	16	8.3
36 to 40	8	3.9	13	6.2	20	10.4
41 to 45	8	3.9	15	7.2	16	8.3
46 to 50	8	3.9	12	5.6	11	5.7
51 to 55	2	1.	3	1.4	13	6.7
56 to 60	7	3.5	2	1.	16	8.3
61 to 65	4	2.	4	2.	9	4.7
66 to 70	3	1.5	4	2.	9	4.7
71 to 75	3	1.5	2	1.	5	2.6
76 to 80	0	0			5	2.6
81 to 85	0	0			5	2.6
86 to 90	1	.5			3	1.6
91 to 95					2	1.
96 to 100					2	1.
Total	204	100	209	100	193	100
Average	25.1		25.5		42.6	
Median	22.6		23.5		40.4	
25 P.	12.9		13.4		26.8	
75 P.	33.5		36.6		57.9	
P.E.	10.3		11.6		15.6	
$P.E._{t.obt.-Av.}$	.7		.8		1.1	

TABLE XXI

PER CENT OF DIVISION COMBINATIONS CORRECTLY ANSWERED GAINED IN THE COURSE OF SIXTY MINUTES OF PRACTICE BY THE THREE GROUPS

Gain, per cent	Group I 20 min.		Group II 10 min.		Group III 2 min.	
	Individuals	Per cents	Individuals	Per cents	Individuals	Per cents
-74 to -60	1	.5				
-59 to -45			3	1.4		
-44 to -30			2	1.		
-29 to -15	3	1.5	2	1.		
-14 to 0	7	3.5	5	2.5	2	1.
1 to 15	8	3.9	5	2.5	3	1.6
16 to 30	16	7.8	18	8.6	8	4.1
31 to 45	26	12.8	24	11.2	9	4.7
46 to 60	42	20.6	20	9.6	15	7.8
61 to 75	27	13.2	31	14.8	27	14.
76 to 90	22	10.8	27	12.9	27	14.
91 to 105	12	5.9	14	6.6	24	12.5
106 to 120	9	4.4	14	6.6	10	5.2
121 to 135	12	6.	6	2.8	12	6.2
136 to 150	4	2.	8	3.8	10	5.2
151 to 165	6	3.	0	0	3	1.6
166 to 180	1	.5	3	1.4	6	3.1
181 to 195	1	.5	5	2.5	2	1.
196 to 210	1	.5	3	1.4	4	2.1
211 to 225	1	.5	3	1.4	6	3.1
226 to 240	1	.5	8	3.8	2	1.
241 to 255	1	.5	1	.5	2	1.
256 to 270			1	.5	3	1.6
271 to 285			0		5	2.6
286 to 300			1	.5	2	1.
301 to 315			0		1	.5
316 to 330	1	.5	0		2	1.
331 to 345			1	.5	1	.5
346 to 360			1	.5	2	1.
361 to 375	1	.5	1	.5	2	1.
376 to 390					2	1.
391 to 405	1	.5	2	1.	1	.5
Total	204	100	209	100	193	100
Median	60		73		94	
25 P.	40		41		67	
75 P.	92		112		147	
P.E.	26		35.3		40	
$P.E._{t.\text{-}obt.\ Av.}$	1.8		2.4		2.9	

Gross Gain

To determine the gross gain of the three groups in division made in the course of 60 minutes of practice, the gross gains in number of combinations worked correctly by each of the

606 children, samples of which may be seen in Table VII, column *16*, were distributed for the three groups in Table XX, and the measures for each group calculated.

The average number of combinations gained in the course of 60 minutes by Group I was 25.1; by Group II, 25.5; and by Group III, 42.6. The corresponding medians were 22.6, 23.5 and 40.4. At their face value these facts show a very decided advantage for the group which did the intervening 45 minutes of practice in 2-minute periods over the groups which did it in 10-minute periods and 20-minute periods. Only 18 per cent of the 20-minute group and 21 per cent of the 10-minute group reached the median of the 2-minute group.

Percentile Gain

The data used in determining the gain per cent for the groups are the individual records of the 606 children in gain per cent, samples of which may be seen in Table VII, column *17*. These gain per cents are distributed (separately) in Table XXI, for the three groups.

Gain in Accuracy

The gross gains in accuracy for the 606 individuals, samples of which may be seen in Table VII, column *18*, were distributed for the three groups in Table XXII and the measures for each group found.

The median gains in accuracy were: for Group I, 2.1 per cent; for Group II, 3.5 per cent; for Group III, 2.3 per cent. These gains are so nearly the same that we can say that in division no advantage was shown in accuracy from one length of practice period over the others. All periods resulted in about an equal absolute gain in accuracy.

Summary

There were 18 3B and 4A classes, containing 606 children in the division experiment. Group I consisted of 6 classes containing 204 children; Group II consisted of 6 classes containing 209 children; and Group III consisted of 6 classes containing 193 children. The groups were made on the basis of length of practice-period in the 40 minutes of practice between the initial 10-minute period and the final 10-minute period, which were the

TABLE XXII

GROSS GAIN IN ACCURACY IN DIVISION, EXPRESSED IN PER CENTS OF ANSWERS THAT WERE CORRECT, MADE IN THE COURSE OF SIXTY MINUTES OF PRACTICE BY THE GROUPS

Gain, per cent	Group I		Group II		Group III	
	20 min.		10 min.		2 min.	
	Individuals	Per cents	Individuals	Per cents	Individuals	Per cents
–45 to –41	1	.5				
–40 to –36			1	.5		
–35 to –31						
–30 to –26			2	1		
–25 to –21	1	.5	1	.5	1	.5
–30 to –16					3	1.6
–15 to –11	6	.3	1	.5	4	2.1
–10 to – 6	15	7.4	6	2.8	16	8.3
– 5 to – 1	36	17.6	38	18.1	39	20.2
0 to 4	82	40.2	70	33.5	59	30.6
5 to 9	28	13.7	43	20.6	31	16.1
10 to 14	17	8.3	13	6.2	22	11.4
15 to 19	3	1.5	10	4.8	13	6.7
20 to 24	4	2	8	3.8	1	.5
25 to 29	5	2.5	5	2.4	1	.5
30 to 34	4	2	3	1.4	1	.5
35 to 39	2	1				
40 to 44			4	2	1	.5
45 to 49			4	2		
50 to 54					1	.5
Total	204	100	209	100	193	100
Median	2.1		3.5		2.3	
25 Percentile	–1.6		–.3		–2.4	
75 Percentile	6.6		8.9		8.2	
P.E.	4.2		4.6		5.3	

same for all three groups. In Group I, these intervening practice periods were 20 minutes each; in Group II, 10 minutes each; in Group III, 2 minutes each. The following is a summary of the important facts concerning the differences in improvement.

MEDIANS OF THE GROUPS OF INDIVIDUALS

	Median Initial Ability	Average Gross Gain	Median Gross Gain	Reliability of Average $P.E._{t.\text{-obt. Av.}}$	Median Gain Per cent	Median Gain in Accuracy
Group I	38.3	25.1	22.6	.7	60	2.1
Group II	32.	25.5	23.5	.8	73	3.5
Group III	40.3	42.6	40.4	1.1	94	2.3

1. The initial ability of Group II was enough below that of the other two groups to demand careful consideration in evaluating the relative gain.

2. The initial accuracy of the groups was so nearly equal that it may be considered a constant factor in the discussion.

3. Group III gained almost twice as many combinations done correctly in the course of the 60 minutes of practice as did the other two groups, whose gains were practically the same.

4. In gross gain in accuracy the three groups were so nearly equal that this too may be considered as almost a constant factor.

The measures given have been computed from the scores of the individuals comprising these groups. The following measures are computed from the scores of the *classes* in the groups. The *median* for each *class* was found, then the *average* of these medians was computed.

AVERAGE OF CLASS MEDIANS

	Average Initial Median Ability of Classes	Reliability of Average: P.E. t.-obt Av.	Average Gross Median of Classes	Reliability of Average: P.E. t.-obt. av.	Average of Median Percentile Gains
Group I	38.4	2.2	20.6	2.	58
Group II	33.4	1.7	25.1	1.5	77
Group III	41.4	4.5	44.7	2.7	114

Using the three methods of computing gross gain, we have:

	AVERAGE GROSS GAIN OF INDIVIDUALS	MEDIAN GROSS GAIN OF INDIVIDUALS	AVERAGE OF MEDIAN GROSS GAINS OF CLASSES
Group I	25.1	22.6	20.6
Group II	25.5	23.5	25.1
Group III	42.6	40.4	44.7

Considering the facts for both addition and division, it appears that, subject to discounts for the inequalities of the groups in initial ability, there is considerable advantage in the short period lengths, when the length is two minutes. The advantage there is noteworthy, since in addition the gain is greater than in the longer-period groups even when their ability was greater in the longer period (Group II); and since in division the gain is so very much greater than in the 20- or 10-minute group.

The facts can be freed from the influence of inequalities in ability at the beginning of practice by comparing only those of equal initial ability. For example, we find in the case of division that those of initial ability 15 averaged in gain, 18.0, 26.3 and 23.0 according as they had practiced in 20-, 10-, or 2-minute periods; those of initial ability 16 averaged in gain 16.4, 15.3 and 41.0 according as they had practiced in 20-, 10-, or 2-minute periods.

Making such calculations for those of each initial ability in division from 5 to 64 and allowing equal weight to each successive set of five successive groups, it appears that on the average the 20-minute, 10-minute, and 2-minute period varieties of practice brought to those of equal initial ability gains in the relation of 100, 110½ and 177.

In the case of addition the same procedure, carried out with those of initial abilities 5, 6, 7, and so on up through 49 gives the following results: According as the practice was in 22½, 15, 6, or 2-minute divisions, it brought to those of equal initial ability gains in the relation of 100, 121, 101 and 146½.

It appears, then, that the superiority of the shortest practice-period length remains when inequalities of initial ability are eliminated. It appears further that the periods of intermediate length have really a greater superiority over the longest period than the results irrespective of differences in initial ability showed. There is a positive relation between initial ability and gross gain. Consequently Group III in addition and Group II in division, which happened to be groups of low initial ability, suffered in the comparison.

The detailed facts for the gains of each variety of initial ability according to the nature of the practice are shown in Tables XXIII and XXIV.

TABLE XXIII

THE RELATION OF LENGTH OF PRACTICE-PERIODS TO GROSS GAIN IN ADDITION

Initial ability. Examples correct	22½ min. practice-periods		15 min. practice-periods		6 min. practice-periods		2 min. practice-periods	
	Average Gross Gain	Number of Cases	Average Gross Gain	Number of Cases	Average Gross Gain	Number of Cases	Average Gross Gain	Number of Cases
0								
1								
2							2	1
3								
4	5	1			14	1		
5	4	2	8	2	4	1	4	1
6							30	4
7	9.3	3	8.	2	2.5	2	14.2	4
8	7.6	3	9.5	2	11.2	5	9.8	5
9	13.4	7			-1.5	2	4.7	4
10	6.6	3	25	1	14.7	11	13.3	3
11	19	1	2	2	13	4		
12	7	9	4	3	17.7	4	9.2	4
13	10.5	4			9.3	6	8.8	5
14	7.3	6	11	3	12.8	8	23.3	3
15	7.6	3	-6	2	14.6	9	19.1	12
16	5.2	6	21.3	3	6.2	5	16.7	8
17	7.5	9	5	3	5	8	14.2	4
18	9.3	3	13.3	4	9 .	21	8	9
19	7.6	11	15	3	4	5	8.2	8
20	7.5	10	6.1	7	10.8	16	11.1	7
21	9.7	9	10.2	5	6	2	17.8	6
22	6.7	4	7	1	12.7	4	9.4	12
23	6.3	7	14	2	6.9	10	4.3	7
24	12.6	8	7.7	4.	9.6	5	20.8	6
25	10	6	21	2	8.1	8	25.8	5
26	7.2	4	28.3	3	7.3	7	11.5	11
27	8.6	8	9.6	3	10.6	7	17.2	5
28	7.5	6	9	1	8.6	8	15	13
29	10.8	4	-3	1		2	11.3	6
30	23.7	6	8.4	5	1.4	5	10.3	4
31	12.4	5	16.3	4	6.8	4	11.9	10
32	14.5	2	22.6	3	3.5	2	25.4	5
33	15	3	37	1	16	3	27.3	6
34	5.6	3	2	1	14.4	5	11.8	4
35	4.8	4	3	1	13.4	7	14.3	3
36	11.8	4	8.8	4	19	1	34.5	2
37			12.8	5	29	1	10.6	8
38	8	1	14	1	22	2	17.6	5

TABLE XXIII—*Continued*

Initial ability. Examples correct	22½ min. practice-periods		15 min. practice-periods		6 min. practice-periods		2 min. practice-periods	
	Average Gross Gain	Number of Cases	Average Gross Gain	Number of Cases	Average Gross Gain	Number of Cases	Average Gross Gain	Number of Cases
39	20	3	61	1			35	2
40	15	1	24	1	10.7	6	33.6	5
41	7	1	3	1			38	3
42	23.5	2	21.5	2	37	1	28.5	2
43	20	1					15	3
44	4.5	2			28	1	8	1
45	27	1					16	1
46	23	1	14	1				
47					16.5	2		
48	33	1			26.5	2		
49			49	2			26.3	3
50	26	1	4	1			34	3
51	39	1	–6	1	19	1		
52	30.5	2	18	1				
53	30	1			29.5	2		
54							53	1
55								
56							57	1
57	29	1						
58							59	1
59	17.5	2					33	1
60– 64	20.5	2	31	1.			12	1
65– 69	28	1			46	1	24	1
70– 74	30.3	3	37	2			43	1
80– 84			23	3				
85– 89	22	1						
90– 94	18	1					18	1
105–109			48	1				
Total		194		102		207		231

TABLE XXIV

THE RELATION OF LENGTH OF PRACTICE-PERIODS TO GROSS GAIN IN DIVISION

Initial ability. Examples correct	20 min. practice-periods		10 min. practice-periods		2 min. practice-periods	
	Average Gross Gain	Number of Cases	Average Gross Gain	Number of Cases	Average Gross Gain	Number of Cases
5	20	1	15.5	2		
6	13	2	12	3		
7	9	1	10.5	2		
8	15.3	3	5	1	24	1
9	4.6	3	11.5	2	12	2
10	10	3	8	1		
11	9.6	3	24	4		
12	12	2	5	3	24	2
13	13.8	5	13.5	2	35.5	2
14	39	2	14.5	2	38	2
15	18	4	26.3	4	23	7
16	16.4	5	15.3	3	41	4
17	15.5	6	13	4	38.2	5
18	2.5	2	40.5	6	44.7	6
19	15	2	11.8	5	41	1
20	-15	1	26.2	9	34.5	2
21	28	2	38.5	2	38.8	5
22	23	2	27.5	8	43.6	5
23	30.6	3	20.5	4	17.7	3
24	14.2	5	30.5	4	74.7	3
25	21.3	3	21.2	6	57	1
26	25	1	13	2	40.2	5
27	10.5	4	17.1	7	20.5	2
28	22.3	4	25.4	8	23.5	2
29	25.3	3	41.5	2	19.5	2
30	23.6	5	19.5	6	38.4	5
31	18.5	4	20	4	40.5	2
32	16.5	2	28.6	3	33.5	2
33	28	2	31.6	3	39.3	4
34	23	2	29	4	33.5	2
35	21	3			44	5
36	19.7	7	31.8	5	44	1
37	22.3	4	31.1	8	46.8	4
38			27.3	3	48.5	5
39	25.3	6	20.8	5	17	2
40	19	4	7.5	2	43	1
41	46	1	30.2	5	46.3	3
42	32	2			39.3	4
43	25.4	8			34	8
44	22.5	2	13	5	45	1

TABLE XXIV—*Continued*

Initial ability. Examples correct	20 min. practice-periods		10 min. practice-periods		2 min. practice-periods	
	Average Gross Gain	Number of Cases	Average Gross Gain	Number of Cases	Average Gross Gain	Number of Cases
45	10.5	2	18	4	64	1
46	6	1	38.3	6	18	3
47	26	1	16.6	3	41.6	3
48	32	4	15.5	2	47.5	2
49	37.5	4	25.8	4	44.3	4
50	17.5	2	44	1	27	1
51	16.6	7	29	7	42.8	4
52	24.3	3	6	1	37.5	2
53	32	3				
54	28.3	3	22	3	67	1
55	19.5	4	19.5	2	31.8	6
56			49	1	44.5	2
57	37	4	27.8	6	40.5	2
58	18.3	3	50	1	27.8	4
59	23.6	3			37.3	3
60	26	2	27	1	83	1
61	25	3	47	1	45.3	3
62	31	3	34	3		
63	41.4	5	49	1	47	1
64	42	3			48	1
65	51	1	63	1	62	1
66	35	1	48.5	2	92	1
67	31	1	48	1		
68	40	1	34.5	2	43	1
69						
70– 74	69.5	2	28.8	4	57.9	9
75– 79	45.4	8	46.6	3	49.6	5
80– 84					63	3
85– 89	64	2	41	1	43	6
90– 94	90	1			74	4
95– 99	59.3	3			77.5	2
100–104	67	1				
105–109					87	2
110–114					62	1
115–119						
120–124					90	1
125–129					72	1
Total		205		210		192

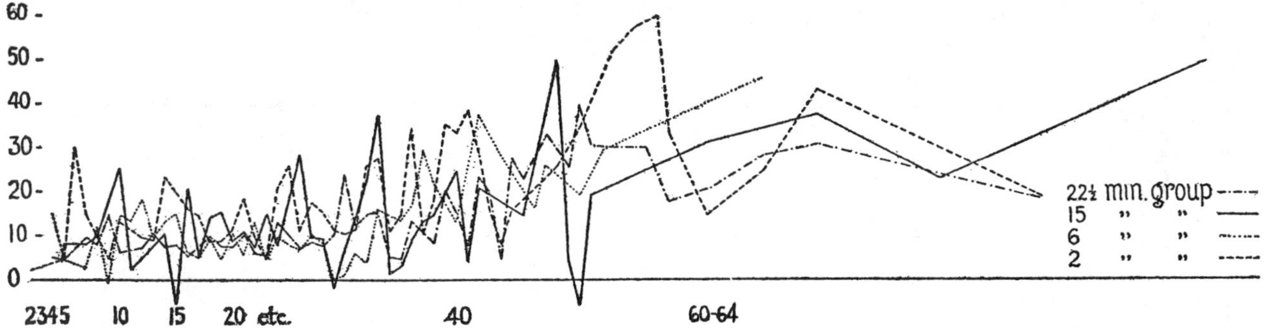

FIG. 1. Improvement in addition in relation to initial ability for each length of practice-period. The horizontal scale is for the initial score; the vertical scale is for gain in examples done correctly.

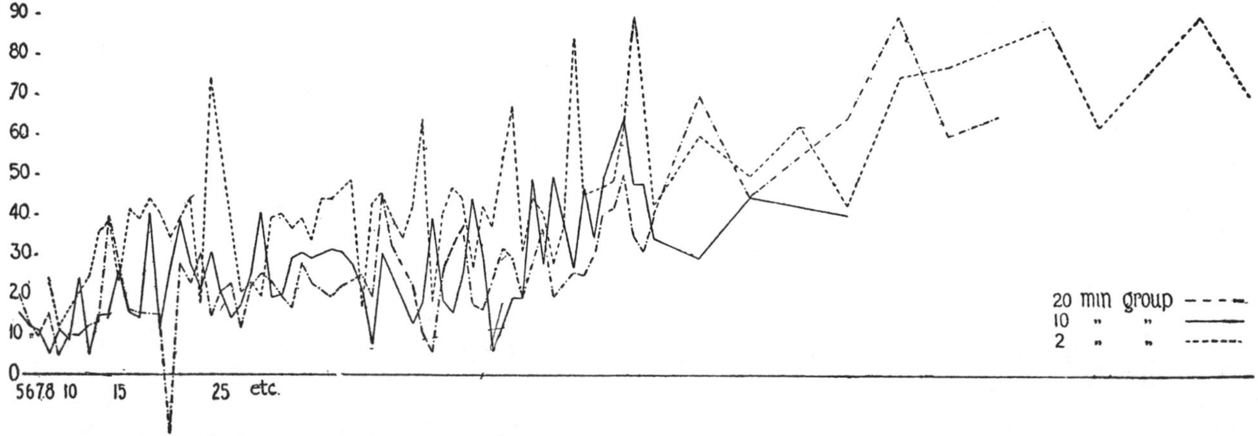

FIG. 2. Improvement in division in relation to initial ability for each length of practice-period. The horizontal scale is for the initial score; the vertical scale is for gain in examples done correctly.

General Summary

While the results of both the addition and division experiment show that the greatest gains were made by the groups which had their practice in the shortest periods, we can not conclude that all of this excess of gain was due to the difference in the length of the practice-period. Other factors probably contributed a share to this extra gain. (1) The groups, working in shorter periods, because of the number of days over which the experiments ran, had greater opportunity during the experiment to profit from the regular school work than other classes. This difference might have been overcome by giving the experiment to all groups in the same number of days, which would necessitate giving more than one of the shorter practice-periods within a day or allowing days to lapse between the longer practice-periods. This is a problem for another study. (2) The groups working in shorter periods had a longer time in which to catch the spirit of the experiment and to become enthusiastic over surpassing their previous performance. They had their records read to them more times and had the incentives to intense effort repeated more often. (3) They also had greater opportunity and incentive to do work outside of the time given to the experiment.

On the other hand the 2-minute group in addition was handicapped by the difficulty of making an absolute gain from one 2-minute period to the next. The median gross gain for this group was 12.6 columns from 23 periods of practice which shows that only those who gained most rapidly could have had the pleasurable effect from day to day of surpassing their previous day's record. Then, too, if they did surpass their previous days record it was by a small amount which did not give the same feeling of success and incentive to effort that a larger gain would have given. This condition did not operate much in the 2-minute period in division where the median gain was 40.4 combinations in 20 periods of practice. In all other periods of practice the gain from day to day was larger and resulted in a more intense feeling of success and greater incentive to effort. This handicap on the 2-minute group, it seems, would offset the extra gains from conditions previously mentioned.

From the administrative side of school-room work, the longer periods of practice can be much more economically managed in that preparation needs be made for them much less often. Perhaps this factor would compensate in a large measure for the disadvantage which the long period seems to have in comparison with short ones.

CHAPTER IV

THE PERMANENCE OF THE PRACTICE EFFECT

In order to measure the permanence of the ability acquired in the course of the experiments that have been described, retention tests were given at the end of the school year in June, 1912, to those classes that had taken their practice during the preceding four months and finished it long enough before the end of June to make a retention test seem profitable.

No one knew that these later tests were to be given after the regular experiment was finished. This makes it reasonable to suppose that about the regular amount of work was done in addition and division from the close of the experiment until the tests for retention were given and that no extra drill was given in anticipation of them. These tests afford data by which to determine to what extent the efficiency acquired during the experiments persisted from the close of the practice in the experiment to the end of June under the normal exercise of performing regular school work.

Retention tests were also given at the beginning of the following school year in September to as many of the same classes as the author had access to. These tests afford data by which to determine to what extent the efficiency shown at the end of June persisted through vacation, during which practice was substantially zero. Immediately following these retention tests in September, the classes were given sufficient practice to restore them approximately to the same efficiency they had attained at the end of their practice the preceding year.

These retention tests correspond in every way to the initial and final practice-periods of the experiment. In addition there were practice-periods of 15 minutes, and in division practice-periods of 10 minutes. Hence, the scores made in them were directly comparable to the scores in both the initial and final periods of the experiment.

In the practice conducted to restore the ability lost during vacation some five-minute practice-periods were used, but the last practice-period given was a 15-minute period in addition and a 10-minute period in division. The retention tests and the practice given in September were conducted by the author in exactly the same manner as the practice of the experiment had been conducted. During some of the 15-minute tests in September some of the children showed signs of weariness that were not noticeable during the original practice. The author felt that this was due to the children's inability to apply themselves to mental work immediately after vacation with the same persistence that they had shown during the experiment, which was conducted at a time when they were accustomed to do continuous mental work. There is no doubt in the author's mind that shorter periods of practice would have been more effective in this practice given during the first two weeks of school, and that the time required for these classes to regain their former efficiency would have been considerably lessened if given in shorter periods. The present experiment affords no data to prove this opinion, but it would be profitable to find out if practice-periods should be shorter in the early weeks of the school term than in the later ones.

This chapter treats, then, three phases of the problem of the permanence of associations: (1) The permanence of well-practiced addition and division associations during a period of time when they were normally used in regular school work. (2) The permanence of the same associations after almost complete disuse during the summer vacation. (3) The permanence of these associations as shown by the amount of practice required to restore them to their former efficiency.

Permanence of Associations Normally Used in School Work

Addition

Nine classes that had taken the addition practice finished the experiment and took the retention test in June at the time indicated below:

CLASS	FINAL PRACTICE PERIOD	RETENTION TEST
XIV	March 28	June 21
XVI	March 28	June 26
XIII	March 29	June 21
XVII	April 18	June 20
VIII	April 19	June 19
IX	April 19	June 18
X	May 9	June 19
XI	May 9	June 19
V	May 27	June 20

The question which we are considering is, What change in ability to add took place in the children of these nine classes in the time intervening between the final practice-period and the retention test near the end of June, a time varying for the different classes from 12 weeks to 3 weeks, in which no special drill was given in arithmetic, but in which regular work in arithmetic was in progress? The scores made in the retention tests were not compared directly with the scores made in the final practice-period of the experiment, but with the scores made in the initial practice-period. So, to answer the question proposed above, we shall first find what gain was made in the retention test near the end of June over the initial test and then compare this gain with that made during the original practice-experiment by these same classes. Any difference found between these two gains will be the change that took place in the interval between the close of practice and the test near the end of June.

Gross Gain: Table XXV gives under (a) the gross gain of these nine classes in the retention test at the end of June over the initial period of the experiment and under (b) the gross gain of these same classes during the original experiment. The median gain in columns added correctly in the retention test near the end of June over the initial practice-period of the experiment was 11 columns, while the median gain of these same classes during the experiment was 10.4 columns. This means that, in the time elapsing between the end of the regular practice of the experiment and the end of June in which only the regular school work in addition was carried on, these classes not only did not lose any of the ability acquired during the experiment, but gained the difference between these two medians or .6 column. The 25 percentile and the 75 percentile together with the median indicate that this gain was distributed all along the curve.

Gain in Accuracy: Table XXVI gives under (a) the gain of

these nine classes in accuracy in the retention test at the end of June over the initial period of the experiment, and under (b) the gain of the same classes in accuracy during the experiment. Between the initial practice-period and the retention test there was a gross gain in accuracy of .2 per cent while during the experiment there was a loss in accuracy of .4 per cent. The difference between these two medians, .6 per cent, is the median gain in accuracy from the end of the original practice to the end of June. This indicates a slight gain in accuracy of performance after the intense practice ceased and normal use of the associations was continued. However, this gain is so small that little weight should be given to it.

TABLE XXV

	(a) Gross Gain in Addition from the Initial Period of the Experiment to the End of June		(b) Gross Gain of the Same Individuals During the Experiment	
Columns gained	Individuals	Per cents	Individuals	Per cents
−15 to −12	4	1.6	1	.4
−11 to − 8	3	1.5	6	2.6
− 7 to − 4	7	2.7	7	2.7
− 3 to 0	14	5.4	25	9.7
1 to 4	35	13.5	40	15.5
5 to 8	39	15.1	35	13.5
9 to 12	44	17.	31	11.9
13 to 16	38	14.3	34	13.1
17 to 20	24	9.3	29	11.2
21 to 24	13	5.1	14	5.4
25 to 28	4	1.6	13	5.1
29 to 32	12	4.7	9	3.5
33 to 36	5	1.9	6	2.6
37 to 40	4	1.6	1	.4
41 to 44	6	2.6	1	.4
45 to 48	2	.8	3	1.5
49 to 52			1	.4
53 to 56	1	.4		
57 to 60			1	.4
61 to 64	1	.4	1	.4
65 to 68	2	.8		
Total	258	100.	258	100.
	Median	11	Median	10.4
	25 Percentile	4.6	25 Percentile	3.1
	75 Percentile	18.1	75 Percentile	17.1
	P.E.	6.8	P.E.	7.

TABLE XXVI

Per cent gained in accuracy	(a) Gain in Accuracy in Addition from Initial Score to the End of June: Individuals	Per cents	(b) Gain in Accuracy of Same Individuals During the Experiment: Individuals	Per cents
−55 to −51			4	1.7
−50 to −46	1	.5		
−45 to −41	3	1.2	1	.5
−40 to −36	1	.5	1	.5
−35 to −31	3	1.2	6	2.4
−30 to −26	5	1.9	9	3.6
−25 to −21	16	6.1	7	2.6
−20 to −16	20	7.8	20	7.8
−15 to −11	22	8.5	25	9.7
−10 to − 6	21	8.2	23	9.
− 5 to −1	32	12.4	32	12.4
0 to 4	36	13.9	42	16.3
5 to 9	37	14.4	28	10.9
10 to 14	20	7.8	22	8.5
15 to 19	12	4.7	17	6.6
20 to 24	16	6.1	8	3.1
25 to 29	8	3.1	2	.8
30 to 34	2	.8	2	.8
35 to 39	1	.5	3	1.2
40 to 44	1	.5	2	.8
45 to 49			3	1.2
50 to 54	1	.5		
55 to 59			1	.5
Total	258	100.	258	100.
	Median	.2	Median	−.4
	25 Percentile	−12.	25 Percentile	−12.2
	75 Percentile	9.	75 Percentile	8.7
	P.E.	10.5	P.E.	10.5

Division

Gross Gain: Only two classes in the division experiment were given the retention test in June. The other classes did not finish the experiment until the closing days of the term. One of these classes finished the experiment June 7, the other June 10. Both took the retention test June 20. Table XXVII gives under (a) the gross gain of these two classes in number of combinations worked correctly in the retention test on June 20 over the initial practice-period of the experiment, and under (b) the gross gain of the same individuals during the experiment. The median gain in the retention test over the initial period was 30.7 combinations while the gain during the experiment was 28.5 combinations. This means that these two classes in an interval of about 12 days not only did not lose any of the proficiency acquired during the intense practice of the experiment but actually gained a median of 2.2 combinations.

Gain in Accuracy: The change in accuracy may be seen in Table XXVIII. From the initial period to the retention test, the median gain was 2.7 per cent, while the median gain during the experiment was 2 per cent. These figures show a median loss of .7 per cent in accuracy in the interval from the close of the experiment to the retention test.

TABLE XXVII

	(a) Gross Gain of Two Classes in Division from Initial Test to End of June		(b) Gross Gain of Same Classes During Experiment	
Combinations gained	Individuals	Per cents	Individuals	Per cents
−4 to 0	1	1.2	1	1.2
1 to 5	1	1.2	2	2.4
6 to 10	3	3.6	6	7.1
11 to 15	2	2.4	3	3.5
16 to 20	7	8.4	10	12.9
21 to 25	16	19.3	13	15.3
26 to 30	11	13.3	11	12.9
31 to 35	15	18.1	13	15.3
36 to 40	10	12.1	4	4.7
41 to 45	6	7.2	9	10.6
46 to 50	3	3.6	5	5.9
51 to 55	2	2.4	1	1.2
56 to 60	1	1.2	1	2.4
61 to 65	1	1.2	1	1.2
66 to 70			1	1.2
71 to 75	1	1.2	2	2.4
76 to 80	1	1.2		
81 to 85				
86 to 90	1	1.2		
91 to 95	1	1.2		
96 to 100				
Total	83	100	83	100
	Median	30.7	Median	28.5
	25 Percentile	22.6	25 Percentile	19.7
	75 Percentile	38.6	75 Percentile	39.6
	P.E.	8.	P.E.	9.9

Summary: For two classes in division, in an interval of about 12 days between the close of the experiment and the retention test, the median gross gain was 2.2 combinations done correctly, and the median loss in accuracy was .7 per cent. Here, as in addition, these classes not only did not recede from the high efficiency reached during the experiment when practice ceased, but they made a slight gain while using these same associations in the performance of regular school work.

Both the addition experiment and the division experiment indicate that these third- and fourth-year children who had improved their proficiency in adding and in doing the division combinations 50 per cent or more by intense practice in a brief space of time, lost none of this recently acquired proficiency so long as they continued to exercise these same functions to the extent demanded by the performance of regular school work.

TABLE XXVIII

	(a) Gross Gain in Accuracy of Two Classes in Division from Initial Test to End of June		(b) Gross Gain of Same Classes During Experiment	
Per cent gained in accuracy	Individuals	Per cents	Individuals	Per cents
-25 to -21			1	1.2
-20 to -16				
-15 to -11	1	1.2		
-10 to -6	4	4.8	2	2.4
-5 to -1	21	25.3	17	20.
0 to 4	31	37.1	35	41.1
5 to 9	16	19.3	14	16.5
10 to 14	4	4.8	8	9.4
15 to 19	2	2.4	3	3.5
20 to 24	3	3.6	1	1.2
25 to 29	1	1.2	2	2.4
30 to 34			1	1.2
45 to 49			1	1.2
Total	83	100.	85	100.
	Median	2.0	Median	2.7
	25 Percentile	-2.8	25 Percentile	-.4
	75 Percentile	6.1	75 Percentile	7.6
	P.E.	4.5	P.E.	4.

Permanence of Associations Through Summer Vacation

Our next problem is to measure the change in ability in addition and division which occurred during the summer vacation, a time when the associations concerned fell into almost complete disuse. This change will be shown by comparing the scores in the retention tests given at the beginning of September with the scores made at the end of June.

Addition

Gross Loss: Five classes containing 152 children that had taken the practice during the latter half of the school year 1911-1912 and the retention test at the end of June 1912 were accessible to the author the following September. School opened September 9. These five classes were given the retention tests September 10 and 11 with one exception. This class had been promoted into three or four other schools and could not be assembled for the test until September 16. Many children did not return in September. Others, of course, had taken their places, but only the records of children who were present at the end of June and on the day of the retention test in September could be used. Table XXIX gives under (a) the gross gain of these five classes in the September retention tests over their standing in the retention tests in June. The median loss in columns added correctly was 7.5 columns. The lower 25 per cent of the group lost 14.1 columns or more and the upper 25 per cent of the group lost 2. columns or less. Only the upper 15 per cent of the group could add as many columns correctly in 15 minutes at the beginning of September as at the end of June.

Loss Per Cent: The loss per cent used in this portion of the discussion was computed by finding what per cent the gross loss was of the number of problems worked correctly in the retention test at the end of June. If a child added 40 columns correctly in the retention test in June and 32 in September his gross loss was 8 columns and his loss per cent was 20, which would appear in the table as —20. These per cents were distributed as shown under (b) in Table XXIX. The median change was a loss of 17 per cent; that is, the group lost during vacation, as shown by the median, 17 per cent of the ability which it had at the end of June. The lower 25 per cent of the individuals lost 32 per cent or more, and the upper 25 per cent lost 6. per cent or less.

Loss in Accuracy: Anyone to whom the method of computing the change in accuracy is not clear may find this discussed under Table I. In Table XXIX, under (c), the gross gains in accuracy are distributed. The median loss was 4.2 per cent. The upper 35 per cent of the class did not lose in accuracy.

TABLE XXIX

GAIN OF FIVE CLASSES IN ADDITION FROM END OF JUNE TO BEGINNING OF SEPTEMBER

(a) Gross Gain			(b) Gain Per Cent			(c) Gain in Accuracy		
Columns Gained	Individuals	Per cent	Gain Per cent	Individuals	Per cent	Per cent Gained	Individuals	Per cent
–47 to –44	2	1.3	–65 to –61	2	1.3	–55 to –51	2	1.3
–43 to –40	1	.7	–60 to –56	5	3.3	–50 to –46	1	.7
–39 to –36	2	1.3	–55 to –51			–45 to –41	3	2.
–35 to –32			–50 to –46	7	4.6	–40 to –36	3	2.
–31 to –28	4	2.6	–45 to –41	4	2.6	–35 to –31	3	2.
–27 to –24	7	4.6	–40 to –36	10	6.6	–30 to –26	2	1.3
–23 to –20	5	3.3	–35 to –31	15	10.	–25 to –21	7	4.6
–19 to –16	12	7.9	–30 to –26	9	5.9	–20 to –16	10	6.6
–15 to –12	14	9.2	–25 to –21	13	8.6	–15 to –11	21	13.8
–11 to –8	29	19.1	–20 to –16	16	10.5	–10 to –6	17	11.2
–7 to –4	29	19.1	–15 to –11	17	11.2	–5 to –1	28	18.4
–3 to –0	24	15.8	–10 to –6	16	10.5	0 to 4	26	16.5
1 to 4	19	12.5	–5 to –1	12	7.9	5 to 9	16	10.5
5 to 8	2	1.3	0 to 4	16	10.5	10 to 14	6	3.9
9 to 12	2	1.3	5 to 9	5	3.3	15 to 19	2	1.3
			10 to 14	1	.7	20 to 24	2	1.3
			15 to 19	2	1.3	25 to 29		
			20 to 24	1	.7	30 to 34	3	2.
			25 to 29	1	.7			
Total	152	100.		152	100.		152	100.
Median		–7.5	Median		–17.	Median		–4.2
25 Percentile		–14.1	25 Percentile		–32.	25 Percentile		–13.8
75 Percentile		–2.	75 Percentile		–6.	75 Percentile		2.8
P.E.		6.1	P.E.		13.	P.E.		8.3

DIVISION

Seven of the classes that had taken the practice during the last half of the preceding school year were accessible in September. These classes were given the retention test on September 10 and 11 except one class which could not be reassembled until September 16. Since all of these classes finished their practice in June, and since only two of them could be given the retention test in June, the September retention score is compared with the final practice-period of the practice in the experiment.

Gross Loss: The gross losses were found by taking the difference between the number of combinations done correctly in the retention test in September and the number done correctly in the final practice-period of the experiment in June. These losses for the 221 individuals comprising these seven classes were distributed in Table XXX under (a). The median loss for the group was 18.3 combinations. About 90 per cent of the group lost during vacation. The upper 25 per cent of the group lost 10 or fewer combinations.

TABLE XXX

GAIN IN SEVEN CLASSES IN DIVISION COMBINATIONS FROM THE END OF JUNE TO SEPTEMBER

(a) Gross Gain			(b) Gain Per Cent			(c) Gain in Accuracy		
Combinations Gained	Individuals	Per cent	Gain Per cent	Individuals	Per cent	Per cent gained in accuracy	Individuals	Per cent
–64 to –60	5	2.3	–70 to –66	1	.5	–55 to –51	1	.5
–59 to –55	1	.5	–65 to –61	2	1.	–50 to –46	1	.5
–54 to –50	3	1.4	–60 to –56	3	1.4	–45 to –41	2	.9
–49 to –45	4	1.8	–55 to –51	2	.9	–40 to –36	1	.5
–44 to –40	7	3.2	–50 to –46	9	4.1	–35 to –31	3	1.4
–39 to –35	9	4.1	–45 to –41	13	5.9	–30 to –26	6	2.7
–34 to –30	22	10.	–40 to –36	11	5.	–25 to –21	8	3.7
–29 to –25	28	12.7	–35 to –31	18	8.2	–20 to –16	12	5.4
–24 to –20	23	10.4	–30 to –26	19	8.6	–15 to –11	19	8.6
–19 to –15	35	15.8	–25 to –21	36	16.3	–10 to –6	42	19.
–14 to –10	34	15.4	–20 to –16	25	11.3	–5 to –1	67	30.3
–9 to –5	16	7.2	–15 to –11	25	11.3	0 to 4	44	19.9
–4 to 0	17	7.7	–10 to –6	15	6.8	5 to 9	11	5.
1 to 5	8	3.7	–5 to –1	14	6.3	10 to 14	3	1.4
6 to 10	8	3.7	0 to 4	11	5.	15 to 19		
11 to 15	1	.5	5 to 9	3	1.4	20 to 24		
			10 to 14	4	1.8	25 to 29		
			15 to 19	5	2.3	30 to 34	1	.5
			20 to 24	1	.5			
			25 to 29	1	.5			
			30 to 34					
			35 to 39	1	.5			
			40 to 44					
			45 to 49	2	1.			
Total	221	100.		221	100.		221	100.

(a)		(b)		(c)	
Median	–18.3	Median	–21.	Median	–4.3
25 Percentile	–28.7	25 Percentile	–32.	25 Percentile	–10.2
75 Percentile	–10.3	75 Percentile	–10.	75 Percentile	–.1
P.E.	9.2	P.E.	11.	P.E.	5.

Loss Per Cent: The loss per cent was computed by finding what per cent the gross loss was of the number of combinations worked correctly in the final practice-period of the experiment. If a child did 75 combinations correctly in the final practice-period in June and 60 combinations in the retention test in September the gross loss was 15 combinations, or 20 per cent. These per cents are distributed in Table XXX under (b). The median loss was 21 per cent.

Loss in Accuracy: The losses in accuracy are given in Table XXX under (c). The median loss was 4.3 per cent.

Both in addition and division there was a great loss in ability during the summer vacation. The loss in division was greater than that in addition. This difference may have been due to the fact that younger children took part in the division work than in the addition or it may have been due to the fact that the children had been working in addition longer than in division.* Both the difference in age of the children and the fact that they had been working in addition longer than in division would afford conditions for causing the addition associations to be more firmly set than the division combinations and hence more likely to persist.

Permanence in Terms of Advance Over the Initial Practice-period of the Original Experiment

The permanence of the improvement may also be measured by the superiority of the scores attained in September to the scores made at the beginning of the original experiment. In presenting these facts I shall present also the gain made from the initial to the final period of the experiment itself. The amount of permanence can then be viewed as the superiority of the September scores to the initial or their superiority to the final scores of the original experiment.

Addition

Gross Gain: The records of five classes of 148 children are available. One of these classes finished the experiment April 14, two May 4, one May 27, and one June 18. The questions to be

*There was not only the excess of 15 minutes of practice in the original experiment, but also the extra June 15 minute retention test. Only two-sevenths of the classes had this June retention test.

answered are, What gain on their initial ability did the pupils of these five classes make from the initial practice-period to the beginning of September? and, How much of their initial ability did they lose from the end of the experiment to the beginning of September? The data for answering these questions are given in Table XXXI. The table shows that in September the individuals in question were still 10.4 columns ahead of their standing at the beginning of the experiment but that from the close of the practice to the beginning of September they lost a median of 6.4 columns.

Change in Accuracy: Similar data concerning accuracy are found in Table XXXII. The median loss in accuracy from

TABLE XXXI

	(a) Gross Gain in Addition from Initial Practice-Period to the Retention Test in September		(b) Gross Gain of the Same Individuals During the Experiment	
Columns gained	Individuals	Per cents	Individuals	Per cents
–19 to –12	1	.7		
–11 to –8	7	4.7	2	1.4
–7 to –4	9	6.1	1	.7
–3 to –0	12	8.1	12	8.1
1 to 4	18	12.2	19	12.8
5 to 8	19	12.8	16	10.8
9 to 12	17	11.5	10	6.8
13 to 16	17	11.5	13	8.8
17 to 20	13	8.8	16	10.8
21 to 24	13	8.8	11	7.4
25 to 28	6	4.1	15	10.1
29 to 32	6	4.1	8	5.4
33 to 36	4	2.7	5	3.4
37 to 40	3	2.	7	4.7
41 to 44			1	.7
45 to 48			3	2.
49 to 52			4	2.7
53 to 56				
57 to 60	1	.7	2	1.4
61 to 64	2	1.4	1	.7
65 to 68			2	1.4
Total	148	100.	148	100.
	Median	10.4	Median	16.8
	25 Percentile	2.	25 Percentile	5.3
	75 Percentile	19.9	25 Percentile	27.2
	P.E.	8.8	P.E.	11.

the initial period to the beginning of September was 4.1 per cent, while the loss in accuracy from the beginning to the end of the original practice was 1.5.

TABLE XXXII

	(a) Gain in Accuracy in Addition from Initial Practice-Period to the Retention Test in September		(b) Gain of Same Individuals in Accuracy During the Experiment	
Per cent gained in accuracy	Individuals	Per cents	Individuals	Per cents
–55 to –51	2	1.4	2	1.4
–50 to –46	2	1.4		
–45 to –41	2	1.4	1	.7
–40 to –36	1	.7		
–35 to –31	2	1.4	3	2.
–30 to –26	8	5.4	4.	2.7
–25 to –21	6	4.	5	3.4
–20 to –16	11	7.4	12	8.1
–15 to –11	17	11.5	13	8.8
–10 to –6	18	12.2	14	9.5
–5 to –1	18	12.2	25	16.9
0 to 4	19	12.8	22	14.9
5 to 9	17	11.5	15	10.1
10 to 14	11	7.4	13	8.8
15 to 19	5	3.4	11	7.4
20 to 24	3	2.	2	1.4
25 to 29	1	.7	4	2.7
30 to 34	2	1.4	1	.7
35 to 39	1	.7		
40 to 44	1	.7		
45 to 49				
50 to 54	1	.7	1	.7
Total	148	100.	148	100.
	Median	–4.1	Median	–1.5
	25 Percentile	–14.6	25 Percentile	–11.7
	75 Percentile	6.	75 Percentile	7.8
	P.E.	10.3	P.E.	9.8

Division

Gross Gain: The gross gain in number of combinations worked correctly from the initial period of the experiment to the beginning of September, and the gross gain of the same individuals during the original experiment, are shown together in

Table XXXIII. The median gain in the test at the beginning of September over the initial period was 17.5 combinations, while the median gain during the experiment was 37.5 combinations. At the beginning of September the group was thus still a median of 17.5 combinations ahead of the median ability shown in the initial period, but had lost a median of 20.0 combinations from the end of practice to the beginning of September.

TABLE XXXIII

	(a) Gross Gain in Division from Initial Practice Period to the Retention Test in September		(b) Gross Gain of the Same Individuals During the Experiment	
Columns gained	Individuals	Per cents	Individuals	Per cents
-14 to -10	3	1.4	1.	.5
-9 to -5	3	1.4	1	.5
-5 to 0	12	5.7	2	.9
1 to 5	28	13.3	2	.9
6 to 10	16	7.6	9	4.3
11 to 15	32	15.2	9	4.3
16 to 20	28	13.3	17	8.
21 to 25	16	7.6	19	9.
26 to 30	21	10.	18	8.6
31 to 35	17	8.	20	9.5
36 to 40	11	5.2	17	8.
41 to 45	8	3.9	18	8.6
46 to 50	8	3.9	14	6.6
51 to 55	2	.9	12	5.7
56 to 60	3	1.5	17	8.
61 to 65	1	.5	10	4.8
66 to 70			9	4.3
71 to 75	1	.5	5	2.4
76 to 80			2	.9
81 to 85			4	1.9
86 to 90			2	.9
91 to 95			1	.5
96 to 100			1	.5
Total	210	100.	210	100.
	Median	17.5	Median	37.5
	25 Percentile	7.5	25 Percentile	23.5
	75 Percentile	30.1	75 Percentile	54.9
	P.E.	11.3	P.E.	15.7

Loss in Accuracy: Table XXXIV shows the facts. The median loss in accuracy from the beginning of the experiment to the beginning of September was .3 per cent while the median gain in accuracy during the experiment was 2.7 per cent. Hence during vacation there was a median loss of 3.0 per cent in accuracy.

It must be remembered that these measures are from one 15-minute test to another, and so do not measure the effect of the disuse of the associations pure and simple. The loss from the last half-minute of the original practice to the first half-minute of the test in September might well be greater. But for practical purposes we wish to know just the fact which is measured here; namely, the change in the ability as shown in a test of ordinary length. Whatever loss is recovered in a minute or two of practice is educationally of no moment.

TABLE XXXIV

	(a) Gain in Accuracy from Initial Practice-Period to the Retention Test in September		(b) Gain of the Same Individuals in Accuracy During Experiment	
Per cent gained	Individuals	Per cents	Individuals	Per cents
–40 to –36			1	.5
–35 to –31				
–30 to –26	2	.9	1	5
–25 to –21	4	1.9		
–20 to –16	7	3.3		
–15 to –11	9	4.3	4	1.9
–10 to –6	19	9.	7	3.3
–5 to –1	61	29.	40	19.
0 to 4	67	31.9	82	39
5 to 9	20	9.6	39	18.5
10 to 14	6	2.9	17	8
15 to 19	10	4.8	7	3.3
20 to 24	2	.9	4	1.9
25 to 29			3	1.4
30 to 34	1	.5	2	.9
35 to 39	2	.9		
40 to 44			2	.9
45 to 49			1	.5
Total	210	100.	210	100.
	Median	–.3	Median	2.7
	25 Percentile	–4.6	25 Percentile	–.5
	75 Percentile	3.6	75 Percentile	7.4
	P.E.	4.1	P.E.	4.

PERMANENCE OF ASSOCIATION IN ADDITION AND DIVISION DURING VACATION AS SHOWN BY THE AMOUNT OF PRACTICE REQUIRED TO RESTORE THESE ASSOCIATIONS TO THEIR PREVIOUS EFFICIENCY.

As soon as possible after the retention tests in September, each class was given sufficient practice in addition and division to restore it approximately to the same efficiency it had attained at the end of its practice the preceding year. Here " approximately " means that the standing of a class, as shown by an average that could be calculated quickly, was within a few units of its standing at the end of the practice experiment.

This practice was conducted under the same conditions as the practice of the experiment except that a practice-period could not always be given to each class on successive school days since there were four classes in addition and six in division to be re-practiced.

ADDITION

Only four classes containing 114 of the children who had taken the practice the preceding year were given practice to make good the loss of vacation.

Time of Practice: These four classes had their retention test and their practice as indicated below:

CLASS	RETENTION PRACTICE TEST	PRACTICE	FINAL PRACTICE
IX	Sept. 10—15 min.	Sept. 16— 5 min.	Sept. 18—15 min.
X	Sept. 12—15 min.		Sept. 17—15 min.
XI	Sept. 12—15 min.		Sept. 17—15 min.
VI	Sept. 11—15 min.	Sept. 17— 5 min. Sept. 18— 5 min. Sept. 19—15 min.	Sept. 25—15 min.

A wide difference in the amount of time required by different classes to regain their former efficiency is apparent. Class VI, which required the greatest amount of time, made a very great gain during the experiment of the previous year and so had a high standard to reach. This may partly account for the extra time required by this class to reach its former efficiency.

The time required to regain this former efficiency was measured in the same way as was the time in the original experiment. While Class IX practiced 35 minutes, its gain was measured for only 27½ minutes. Its rate of adding during the last 15-minute period was the rate of the mid-point of that period, or at the end of 7½ minutes; hence the actual time required for this class to regain its former efficiency was the 15 minutes of practice in the retention test of September 10, plus the 5 minutes on September 16, plus the 7½ minutes of the final practice-period, or 27½ minutes. The time for the other classes was calculated in the same way. Accordingly Class X practiced 22½ minutes; Class XI, 22½ minutes; and Class VI, 47½ minutes. The average time required to regain approximately the efficiency shown at the end of the practice of the preceding year was 30 minutes for the four classes. In other words, 30 minutes of practice were required to bring this group approximately to the efficiency which it reached as a result of 60 minutes of practice the preceding year.

COMPARATIVE STANDING OF THESE FOUR CLASSES AT THE END OF THE EXPERIMENT AND AT THE END OF SEPTEMBER PRACTICE

Gross Gain: Table XXXV gives under (a) the gross gain of the pupils of these four classes at the end of practice in September over their standing in the initial period of the original experiment, and under (b) the gross gain of these same individuals during the original experiment.

A comparison of these two gains will show how nearly the 30 minutes of practice restored the group to its former efficiency. The median gain at the end of the 30 minutes of practice was 12.5 columns, while the median gain for the same group during the experiment was 14.3 columns. This means that 30 minutes of practice restored the median ability of the group to within 1.8 columns of the median ability of the group at the end of the experiment of the previous year.

Accuracy: Table XXXVI shows that the median gain in accuracy for this group at the end of 30 minutes of practice over its accuracy in the initial practice-period was —2 per cent. The loss of the group during the experiment was 3.1 per cent. Hence, so far as accuracy of performance is concerned, the condition of the group was practically the same at the end of the 30 minutes of practice in September as at the end of the experiment.

The facts of Tables XXXV and XXXVI, in other words, are: 114 individuals who had taken part in the addition practice of the preceding year practiced again in September until they reached approximately the ability they had reached at the end of the experiment the preceding year. The average duration time of practice was 30 minutes, at the end of which time the group was within 1.8 columns of the standing which it had reached at the end of the 60 minutes of practice in the experiment of the previous year. The accuracy of performance was practically equal. Hence more than one-half as much practice was required in September to regain the standing reached in the experiment before vacation as was required at that time to reach it.

TABLE XXXV

	(a) Gross Gain in Addition from the Initial Practice-Period of the Experiment to the Last Practice in September		(b) Gross Gain of the Same Individuals During the Experiment	
Columns gained	Individuals	Per cents	Individuals	Per cents
-11 to - 8	4	3.5	2	1.7
-7 to -4	4	3.5	2	1 7
-3 to 0	7	6.1	9	7.9
1 to 4	14	12.3	16	14.
5 to 8	16	14.	13	11.4
9 to 12	12	10.5	10	8.7
13 to 16	14	12.2	11	9.6
17 to 20	10	8.7	10	8.7
21 to 24	5	4.4	8	7
25 to 28	6	5.3	8	7
29 to 32	7	6.1	4	3.5
33 to 36	3	2.6	4	3.5
37 to 40	3	2.6	6	5.3
41 to 44			1	.9
45 to 48	5	4.4	2	1.7
49 to 52			3	2.6
53 to 56	1	.9		
57 to 60	1	.9	2	1.7
61 to 64	1	.9	1	.9
65 to 68	1	.9	2	1.7
Total	114	100.	114	100.
	Median	12.5	Median	14.3
	25 Percentile	4.4	25 Percentile	4.4
	75 Percentile	24.1	75 Percentile	26.8
	P.E.	9.9	P.E.	10.2

TABLE XXXVI

Per cent gained	(a) Gain in Accuracy in Addition from the Initial Practice-Period of the Experiment to the Last Period in September		(b) Gain in Accuracy of the Same Individuals During the Experiment	
	Individuals	Per cents	Individuals	Per cents
-55 to -51	1	.9	3	2.6
-50 to -46	1	.9		
-45 to -41	2	1.8	1	.9
-40 to -36	3	2.6		
-35 to -31	1	.9	4	3.5
-30 to -26	5	4.4	4	3.5
-25 to -21	3	2.6	5	4.4
-20 to -16	8	7	10	8.7
-15 to -11	10	8.7	11	9.6
-10 to -6	16	14	10	8.7
-5 to -1	10	8.7	19	16.6
0 to 4	15	12.1	16	14
5 to 9	10	8.7	10	8.7
10 to 14	6	5.3	7	6.1
15 to 19	9	7.9	7	6.1
20 to 24	7	6.1	2	1.8
25 to 29	4	3.5	3	2.6
30 to 34	2	1.7	1	.9
35 to 39	1	.9	1	.9
40 to 44				
Total	114	100.	114	100.
	Median	-2.	Median	-3.1
	25 Percentile	-13.3	25 Percentile	-14.8
	75 Percentile	9.9	75 Percentile	5.8
	P.E.	11.6	P.E.	10.3

Division

Time of Practice: Only six classes, containing 163 children who had taken the practice the preceding year, could be repracticed in September. These six classes had their retention test and their practice as follows:

Class	Retention Practice-test	Practice	Final Practice-period	Amount
XII	Sept. 10—10 min.	Sept. 16— 5 min. Sept. 18—10 min.	Sept. 19—10 min.	30*
X	Sept. 10—10 min.	Sept. 16— 5 min. Sept. 18—10 min. Sept. 23—10 min.	Sept. 25—10 min.	40
XI	Sept. 10—10 min.	Sept. 16— 5 min.	Sept. 18—10 min.	20
VII	Sept. 10—10 min.	Sept. 16— 5 min. Sept. 18—10 min.	Sept. 19—10 min.	30
VI	Sept. 11—10 min.	Sept. 17— 5 min. Sept. 18—10 min.	Sept. 19—10 min.	30
XIX	Sept. 11—10 min.	Sept. 16— 5 min. Sept. 20—10 min.	Sept. 25—10 min.	30

The average time required for these classes to regain approximately their former ability was 30 minutes.* In order to give an exact basis for comparing the standing of this group at the end of 30 minutes of practice in September with its standing after 50 minutes of practice in the experiment of the previous year, the facts of Tables XXXVII and XXXVIII are presented.

COMPARATIVE STANDING OF THESE SIX CLASSES AT THE END OF THE EXPERIMENT AND AT THE END OF PRACTICE IN SEPTEMBER

Gross Gain: Table XXXVII shows that the median gain from the initial period of the original experiment to the end of the 30 minutes of practice in September was 30.3 combinations, while, during the original experiment itself, the median gain was 34.2 combinations. The 30 minutes of practice lacked 3.9 combinations of restoring the ability of this group to the ability it reached at the end of 50 minutes of practice the preceding spring.

Accuracy: Table XXXVIII shows that the median gain in accuracy at the end of the 30 minutes of practice over the accuracy of the initial performance in the original experiment was 2.6 per cent; while during the experiment of the previous year a gain of 3.0 per cent was made.

In other words Tables XXXVII and XXXVIII report as follows: 163 individuals who had taken part in the division practice of the preceding year practiced again in September until they reached approximately the ability reached at the end

* Here, too, the rate for the last 10 minutes was the rate at the mid-point of the period or at the end of 5 minutes. Hence only 5 minutes of the last 10 minutes are included in the time required to restore the former ability.

of the experiment the preceding year. The average duration of practice was 30 minutes, at the end of which time the group was within 3.9 combinations of the standing which it had reached at the end of the 50 minutes of practice in the experiment of the previous year. There was also a difference of .4 per cent in accuracy of performance. Hence more than three-fifths as much practice was required in September to regain the standing reached in the experiment of the previous year as was required at that time to reach it.

TABLE XXXVII

	(a) Gross Gain in Division from the Initial Period of the Experiment to the Last Practice in September		(b) Gross Gain of the Same Individuals During the Experiment	
Combinations gained	Individuals	Per cents	Individuals	Per cents
–19 to –15	1	.6		
–14 to –10	1	.6	1	.6
–9 to –5	2	1.2		
–4 to 0	4	2.5	2	1.2
1 to 5	5	3.	2	1.2
6 to 10	8	4.9	8	4.9
11 to 15	11	6.8	7	4.3
16 to 20	16	9.8	17	10.4
21 to 25	18	11.	19	11.6
26 to 30	16	9.8	13	8.
31 to 35	11	6.8	17	10.4
36 to 40	15	9.2	14	8.6
41 to 45	13	8.	15	9.2
46 to 50	8	4.9	11	6.8
51 to 55	10	6.1	9	5.5
56 to 60	5	3.	11	6.8
61 to 65	6	3.7	5	3.
66 to 70	3	1.9	4	2.5
71 to 75	2	1.2	4	2.5
76 to 80	4	2.5		
81 to 85	1	.6	2	1.2
86 to 90				
91 to 95	2	1.2	1	.6
96 to 100	1	.6	1	.6
Total	163	100.	163	100.
	Median	30.3	Median	34.2
	25 Percentile	18.2	25 Percentile	21.5
	75 Percentile	46.3	75 Percentile	48.8
	P.E.	14.	P.E.	13.7

TABLE XXXVIII

	(a) Gain in Accuracy in Division from the Initial Period of the Experiment to the Last Period in September		(b) Gain in Accuracy of the Same Individuals During the Experiment	
Per cent gained	Individuals	Per cents	Individuals	Per cents
–45 to –41	1	.6		
–40 to –35	2	1.2	1	.6
–33 to –31				
–30 to –26	1	.6	2	1.2
–25 to –21				
–20 to –16	1	.6		
–15 to– 11	2	1.2	2	1.2
–10 to –6	10	6.1	6	3.7
–5 to –1	30	18.4	30	18.4
0 to 4	56	34.3	58	35.5
5 to 9	26	16.	31	19.
10 to 14	14	8.6	14	8.6
15 to 19	9	5.5	7	4.3
20 to 24	3	1.9	4	2.5
25 to 29	4	2.5	3	1.9
30 to 34	1	.6	2	1.2
35 to 39	2	1.2		
40 to 44			2	1.2
45 to 49	1	.6	1	.6
Total	163	100.	163	100.
	Median	2.6	Median	3.
	25 Percentile	–1.5	25 Percentile	–.5
	75 Percentile	8.2	75 Percentile	8.2
	P.E.	4.9	P.E.	4.4

APPENDIX I

THE USE OF THE METHOD OF THE PRACTICE EXPERIMENT IN TEACHING HANDWRITING AND SPELLING

Handwriting: The method of the practice experiment was applied by the author to writing in five fourth-year classes. Pages of Robinson Crusoe were printed to serve as material for the classes to copy. Uniform penholders and pens were supplied in each class. The regular school paper, of good quality, white and ruled, was used for the writing. The children were supplied with more material than they could copy in the time allowed. They were told to write as rapidly and as well as they could. About equal stress was put upon quantity and quality. Proper position was encouraged.

Class							
Class I	practiced	3	minutes a day	for	16 successive	school	days.
Class II	"	3	"	"	9 alternate	"	"
Class III	"	1	"	"	12 successive	"	"
Class IV	"	1	"	"	9 alternate	"	"
Class V	"	6	"	"	12 successive	"	"

The papers were not scored during the practice so the children had no definite measure of the quantity and quality from one day to the next as they had in arithmetic. This omission made the method radically different, since the incentive from knowing their exact progress was lacking. However, the children were always asked if they had written more and better than on the previous day. The author and the grade teacher observed the writing and urged that improvement in quality was as desirable as improvement in quantity.

The first and last days' papers were scored for the number of letters written. They were also scored for quality by six different judges, by means of Thorndike's Scale for Handwriting, and the average judgment for individuals and class was found. There was a decided gain in quantity in all the classes and a gain in quality in two classes. The following table gives (1) the average number of letters written per minute in the initial performance, (2) the average gain per minute in number of letters in the last day's performance over the first, and (3) the average change of quality expressed in steps of Thorndike's scale.

	Average Number of Letters Written per Minute in Initial Performance	Average Number of Letters Gained per Minute	Average Change in Steps of Thorndike's Scale
Class I	35	7	+ .02
Class II	24	22	—1.16
Class III	22	32	+ .17
Class IV	43	21	—1.53
Class V	17	16	— .9

It is easy to interpret the results where there was a gain both in quantity and in quality as was the case in Classes I and III, the first of which gained 20 per cent in quantity and slightly in quality, the second of which gained 145 per cent in quantity and slightly in quality. Just what merit attaches to the performance where there was a very great gain in quantity, but a loss in quality, remains for future interpretation.

The results suggest that it is possible by means of the practice experiment, to increase speed in writing in the case of fourth-year school children very much and at the same time increase the quality. Had there been a constant measure of quality to act as a check upon quantity, no doubt the classes that lost in quality might have transferred part of the effort from speed to quality, thus maintaining or improving the quality and at the same time increasing their speed.

An interesting problem for future experiment would be one similar to that outlined above with the added feature of having copies of Thorndike's scale hanging in the room readily accessible to children by which they might measure at any time the quality of their own writing. This would serve as an incentive to improvement in quality along with speed.

Spelling: Five fourth-year classes were given work in spelling with the practice experiment as a method. A list of words was used, each one of which 60 per cent of the fourth-year children in a large school of high grade had missed in a certain test. There were about 100 words in the list. These words were placed on cards, one on each card, which were shuffled until they were promiscuously arranged. Then these words were dealt out in three piles until 30 words were placed in each pile. These groups of 30 words were then recorded for use. Then these 90 words were shuffled as before. Next they were dealt out in six piles until there were 12 words in each pile. These lists were recorded for use. Then these 90 words were shuffled again as before. This time they were dealt out in fifteen piles of 6 words each. These lists of 6 words were recorded.

The groups of 6 words and 12 words were printed on large cards to be hung in the front of the room during the study period. It was found impracticable to put the list of 30 words on such a card, so they were typewritten, a copy for each child.

Class I had 15 minutes to study 30 words on each of 3 successive school days

Class II had 6 minutes to study 12 words on each of 6 successive school days

Class III had 6 minutes to study 12 words on each of 6 alternate school days

Class IV had 3 minutes to study 6 words on each of 13 successive school days

Class V had 3 minutes to study 6 words on each of 9 alternate school days

The children were told that there were words on the cards which they should study when the cards were turned at a given signal. They were told how much time they would have for study. They were told to study in any way that they choose, except that they should not take up their pencils during the study. This was done to prevent any attempt at writing words so that they could be used later. They were given no further instructions as to how they should study, except that they should try to learn every word.

Before the study was begun, papers and pencils were passed out to be ready for use immediately after the study ended.

At a given signal the chart or typewritten paper was turned for study. They studied diligently until the chart was turned or the signal to turn their typewritten papers face down was given. Then the words on the chart were pronounced by the experimenter in a promiscuous order at the rate of about 4 words per minute, and written by the pupils.

The papers were marked before the next day's practice. Before beginning the following day's study the scores were read, and good records commended. All were urged to try to beat their previous day's record. This was continued for the number of days indicated above.

The following is the record comprising (1) the percentage of words spelled correctly during the entire experiment, (2) the percentage of words spelled correctly in the first two tests, and (3) the percentage of words spelled correctly in the last two tests.

	Per Cent of Words Spelled Correctly in Entire List	Per Cent of Words Spelled Correctly in First Two Tests	Per Cent of Words Spelled Correctly in Last Two Tests
Class I	82	84	82
Class II	91	87	92
Class III	87	81	93
Class IV	77	88	70
Class V	93	95	90

The results indicate, as have other studies of spelling by Cornman and Rice, that spelling is not amenable to method. Class II and Class III showed improvement in the last two tests over the first two, but the other three classes showed a loss.

APPENDIX II

Sample of Addition Sheets

Reduced to ⅔ of the original size

7	4	9	2	3	7	2	7	8	8	3	5	6	9	8	7
8	2	3	9	7	4	6	5	6	6	7	9	2	4	6	3
6	3	6	6	8	8	3	2	3	3	3	5	7	7	3	8
5	2	6	3	5	3	8	7	6	8	8	6	5	8	6	5
6	9	5	7	3	3	3	5	9	4	2	8	2	8	9	3
5	5	8	2	2	7	9	8	5	5	6	2	9	2	5	6
3	8	4	9	8	5	9	6	2	7	9	9	8	9	2	6
5	7	8	6	3	7	5	4	9	4	3	7	4	2	9	7
7	5	3	3	9	8	2	8	7	9	6	7	2	7	7	6
7	5	2	8	8	5	5	3	2	3	8	3	7	6	2	9

8	9	6	4	7	3	6	2	4	5	4	4	4	5	6	7
7	4	7	6	4	8	5	5	7	9	7	8	6	7	2	9
5	4	6	5	6	5	8	6	5	4	4	3	3	9	9	6
8	7	3	4	9	7	7	8	9	8	5	9	5	2	6	4
9	6	9	9	7	9	7	3	6	2	7	6	8	4	5	9
2	5	6	2	8	4	4	8	8	9	6	2	3	8	6	7
6	9	8	7	3	4	9	8	5	6	6	4	5	3	3	5
9	4	2	2	4	8	8	7	8	3	8	9	6	9	5	9
4	4	4	9	5	9	3	3	3	8	9	2	2	8	9	8
3	8	7	5	4	6	6	8	2	4	4	8	6	2	7	6

7	7	5	9	2	2	3	8	8	5	6	2	3	9	6	7
5	5	8	9	7	5	5	2	4	7	6	7	9	3	7	5
8	2	3	8	9	8	7	8	2	6	7	9	8	3	9	9
7	7	6	6	4	8	2	9	5	8	4	7	7	8	5	2
2	5	4	9	4	7	6	4	9	6	4	2	3	4	7	5
8	8	7	8	6	7	8	6	7	3	5	9	2	2	6	3
9	6	6	6	6	2	7	2	5	3	4	6	3	6	4	4
2	4	5	3	3	9	9	7	9	7	7	3	4	9	5	8
6	8	9	2	5	6	4	9	2	5	9	5	7	5	4	9
8	3	6	6	8	9	9	5	3	4	6	7	9	6	9	9

Sample of Division Sheets

Reduced to about ⅔ of original size

20=	5s[1]		31=	7s[3] and	r.	22=	6s[5] and	r.
56=	9s and	r.	83=	9s and	r.	53=	6s and	r.
30=	7s and	r.	21=	7s		33=	4s and	r.
89=	9s and	r.	54=	8s and	r.	77=	8s and	r.
20=	8s and	r.	32=	4s		22=	9s and	r.
56=	6s and	r.	80=	9s and	r.	52=	7s and	r.
31=	4s and	r.	22=	3s and	r.	33=	7s and	r.
86=	9s and	r.	53=	9s and	r.	75=	9s and	r.
21=	4s and	r.	32=	7s and	r.	23=	5s and	r.
55=	7s and	r.	78=	9s and	r.	51=	8s and	r.
34=	4s and	r.	24=	7s and	r.	36=	7s and	r.
74=	8s and	r.	50=	6s and	r.	67=	9s and	r.
23=	8s and	r.	35=	7s		25=	9s and	r.
45=	9s		69=	9s and	r.	48=	6s	
34=	7s and	r.	25=	3s and	r.	38=	7s and	r.

72=	9s[2]		49=	7s[4]		66=	9s[6] and	r.
24=	4s		36=	4s		26=	5s and	r.
50=	9s and	r.	68=	9s and	r.	47=	8s and	r.
35=	4s and	r.	25=	6s and	r.	39=	4s and	r.
71=	8s and	r.	48=	9s and	r.	65=	9s and	r.
26=	8s and	r.	62=	9s and	r.	38=	4s and	r.
47=	5s and	r.	28=	3s and	r.	59=	9s and	r.
39=	7s and	r.	44=	9s and	r.	29=	5s and	r.
64=	9s and	r.	40=	5s		42=	7s	
27=	4s and	r.	61=	9s and	r.	41=	8s and	r.
46=	7s and	r.	28=	6s and	r.	59=	6s and	r.
37=	4s and	r.	44=	6s and	r.	29=	8s and	r.
63=	9s		41=	5s and	r.	43=	5s and	r
27=	7s and	r.	60=	9s and	r.	57=	8s and	r.
45=	7s and	r.	28=	9s and	r.	58=	7s and	r.
37=	7s and	r.	43=	8s and	r.	31=	4s and	r.